BY THE SAME AUTHOR

Understanding Biology through Evolution, 3rd edition

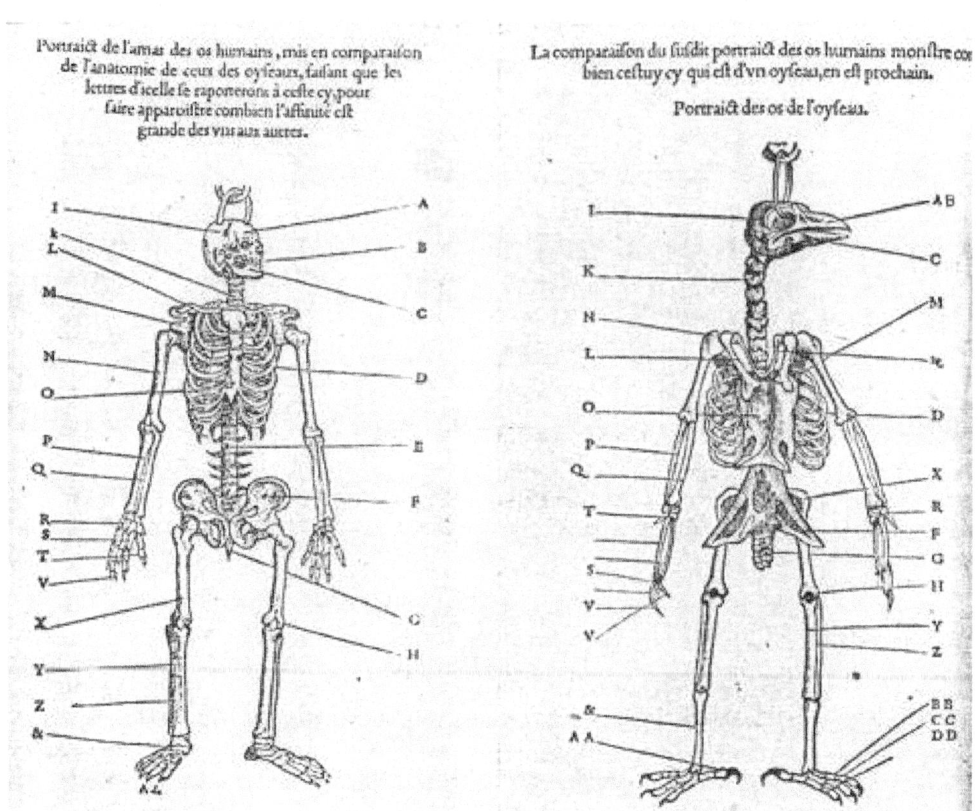

The frontispiece shows the woodcut of a human and a bird skeleton next to each other from a treatise entitled L 'Histoire de la Nature des Oyseaux (The Natural History of Birds), *written by the French naturalist Pierre Belon and published in 1555. The bones of the bird and of the human were similarly labeled, indicating what were formerly called "true affinities" and what today are called homologies. These are structures that develop embryologically in two different species in the same way from the same tissue and are really the same structure, even though they may not look alike or even perform the same function. Belon was the first to notice homologies between specific bones in vertebrates from fish to mammals.*

Over the years, other anatomists discovered other examples of homologous structures, and invariably they "explained" each as being part of a basic body plan, or archetype, in the Creator's mind. Eventually, the hypothesis that these similarities are the result of descent with modification from a common ancestor was proposed but was at first almost universally condemned, just as the earlier Copernican hypothesis of heliocentrism had been. Both theories were considered atheistic, which they are not, since neither concept in any way denies (or confirms) the existence of a Creator.

Understanding

Human Anatomy

Through Evolution

SECOND EDITION

BRUCE D. OLSEN
Los Angeles Trade Technical College

Lulu Press, Inc.
860 Aviation Parkway, Suite 300
Morrisville, North Carolina 27560
http://www.lulu.com

FRONT COVER ILLUSTRATION: The Edwin Smith Papyrus (ca. 1700 B.C.E.), one of the oldest of all
known medical papyri, discusses the cranial sutures, the meninges (coverings of the brain), the
external surface of the brain, the spinal cord, and the cerebrospinal fluid for the first time in recorded
history. Plate 6 and 7 of the papyrus, pictured here, discuss facial trauma. Retrieved from
<http://commons.wikimedia.org/wiki/File:Edwin_Smith_Papyrus_v2.jpg>. ILLUSTRATIONS WITHIN
THIS WORK: Anatomical engravings where indicated are from Henry Gray, *Anatomy of the Human
Body* 20th ed. (Philadelphia: Lea & Febiger, 1918).

In the first place we must look to the constituent parts of animals…And, of course, man is the animal with which we are all of us the most familiar.

– Aristotle, from *Historia Animalium* (ca. 350 B.C.E.)

The fact is that as a natural historian I have yet to find any characteristics which enable man to be distinguished on scientific principles from an ape.

– Carolus (Carl) Linnaeus (von Linné), from the foreword to *Fauna Svecica* (1746)

Man still bears in his bodily frame the indelible stamp of his lowly origin.

– Charles Darwin, from *The Descent of Man* (1871)

<u>ABOUT THE AUTHOR</u>

Bruce Olsen is currently an adjunct professor of life sciences at Los Angeles Trade-Technical College. He received his M.A. in biology from the University of California at Los Angeles in 1998 and his B.A. in biology and chemistry from Case Western Reserve University in 1974. He is a member of several professional organizations, including the American Association for the Advancement of Science, the Society for the Study of Evolution, the American Institute of Biological Sciences, the National Center for Science Education, and the Text and Academic Authors Association.

<u>DEDICATION</u>

To Robert Weinstock, M.D.

| **Topic** | # CONTENTS | **Page** |

PREFACE TO THE SECOND EDITION

Certain aspects of human anatomy, physiology, and pathology make little sense without recognizing that **the human body has a history**. There are a thousand flaws and frailties that make humans vulnerable to disease. Why do we suffer from inguinal hernias and herniated intervertebral discs? If the eye, heart, and brain are such elegant mechanisms, why do we suffer from nearsightedness, heart attacks, and Alzheimer's disease? If our immune system is so sophisticated that it can recognize and attack molecules that have never appeared before on Earth, why do we nevertheless get pneumonia? Why does the immune system sometimes attack our own tissues, causing multiple sclerosis, rheumatic fever, arthritis, diabetes, and lupus erythematosus? As Randolph Nesse and George Williams have written:

> The design of our bodies is simultaneously extraordinarily precise and unbelievably slipshod. It is as if the best engineers in the universe took every seventh day off and turned the work over to bumbling amateurs.[1]

The answer to all of these questions is that the human body has **evolved** through natural selection. In evolutionary terms, we harbor flaws because natural selection, the process that molds our genetically-controlled traits, produces a body plan that allows individuals to survive only long enough to reproduce successfully. Natural selection does not produce perfection or endless good health[2], as Charles Darwin acknowledged:

> Nor ought we to marvel if all the contrivances in nature be not, as far as we can judge, absolutely perfect; and if some of them be abhorrent to our ideas of fitness. We need not marvel at the sting of the bee causing the bee's own death; at drones being produced in such vast numbers for one single act, and being then slaughtered by their sterile sisters; at the astonishing waste of pollen by our fir trees; at the instinctive hatred of the queen bee for her own fertile daughters; at ichneumonidæ feeding within the live bodies of caterpillars; and at other such cases. The wonder indeed is, on the theory of natural selection, that more cases of the want of absolute perfection have not been observed.[3]

Unfortunately, many people, particularly in the United States, still refuse to accept evolution as a fact of life, and what Darwin wrote in his treatise on the evolution of humans is still true today:

> I am aware that the conclusions arrived at in this work will be denounced by some as highly irreligious; but he who denounces them is bound to shew why it is more irreligious to explain the origin of man as a distinct species by descent from some lower form, through the laws of variation and natural selection, than to explain the birth of the individual through the laws of ordinary reproduction.[4]

In other words, each of us began life as a single, fertilized egg, smaller than the period at the end of this sentence, about the size of an amoeba. Few people, if any, find this incompatible with their religious beliefs, or think it implies that a human is nothing more than a large amoeba. Why, then, is the evolution of humans from an ancestral primate any more difficult to accept?

It is all too apparent that undergraduate students often have this same denial about the fact of Darwinian evolution, even in the face of an ever-expanding, already overwhelming body of evidence that supports it. Biologist James Colbert of Iowa State University at Ames has polled students in his introductory biology class for the past three years, asking them if they believe God created humans within the past 10,000 years. In the fall of 2005, he found that 32 percent of a class of 150 students said they did, an alarming proportion "when one considers that these students are academically among the upper half of high school graduates, and they are students choosing to major in a life science"– often to become doctors or veterinarians. "It's very common to see students who simply can't believe humans evolved from apes," he says.[5]

But understanding the evolution of the design flaws of the human body does not just satisfy intellectual curiosity; it is also a vital tool in the quest to prevent and treat disease. For example, a recent editorial in *Science* entitled "Medicine Needs Evolution" states:

> The incorrect idea that selection reliably shapes a happy coexistence of hosts and pathogens persists, despite evidence for the evolution of increased virulence when disease transmission occurs through vectors such as insects, needles, or clinicians' hands. There is growing recognition that cough, fever, and diarrhea are useful responses shaped by natural selection, but knowing when it is safe to block them will require studies grounded in an understanding of how selection shaped the systems that regulate such defenses and the compromises that had to be struck.[6]

This is why students of human anatomy, who often pursue careers in medical occupations, need to know about natural selection and human evolution. The editorial quoted above recommends that evolution be incorporated "into every relevant high school, undergraduate, and graduate course." I agree.

I want to conclude on a personal note. Sometimes I ask myself, why do I continue to write, since the hours are long and the financial reward short? Perhaps the German physicist Max Born (1882-1970) said it best: "To present a scientific subject in an attractive and stimulating manner is an artistic task, similar to that of a novelist or even a dramatic writer. The same holds for writing textbooks."[7] And for me, art is its own reward.

Every effort has been made to present the subject matter in this work accurately and clearly. The author will appreciate readers calling to his attention any inaccuracies and inconsistencies that may have escaped his notice.

Bruce D. Olsen
34° 5' 24.04" N, 118° 22' 42.26" W
olsenbd@lattc.edu

REFERENCES.

[1] Randolph M. Nesse and G.C. Williams, *Why We Get Sick: The New Science of Darwinian Medicine* (New York, NY: Vintage Books, 1994) 5.

[2] S. Jay Olshansky, et al., "If humans were built to last," *Scientific American* 284(3): 51 (2001).

[3] Charles Darwin, *The Origin of Species* 1/e (London, England: Penguin Books Ltd., 1985; originally published: London, England: John Murray, 1859) 445.

[4] Charles Darwin, *The Descent of Man* 2/e (Amherst, New York: Prometheus Books, 1998; originally published: New York: Crowell, 1874) 636.

[5] Constance Holden, "Darwin's place on campus is secure – But not supreme," *Science* 311(5762): 769-71 (Feb. 10, 2006).

[6] Randolph M. Nesse, et al, "Medicine needs evolution," *Science* 311(5764): 1071 (Feb. 24, 2006).

[7] W.F. Bynum and Roy Porter, eds., *Oxford Dictionary of Scientific Quotations* (Oxford, UK: Oxford Univ. Press, 2005) 77.

INTRODUCTION: ANATOMY FROM IMHOTEP TO VESALIUS

> In the history of human anatomy, two figures
> stand out: Claudius Galenus, also known as
> Galen of Pergamum, and Andreas Vesalius.

Ancient Egyptian hieroglyphics are convincing evidence that at a very early date, Egyptians knew considerable gross human and animal anatomy. Many of the signs that represent consonants, vowels, things, and concepts are well-reproduced animals and parts of anatomy. Graphical reproduction of anatomical parts requires knowledge of anatomy. This is also true of sculpture and embalming, which the ancient Egyptians also practiced.

Imhotep in the third dynasty (28th cent. B.C.E.) is credited as the founder of ancient Egyptian medicine and as the original author of the Edwin Smith medical papyrus. This text begins by addressing injuries to the head, and continues with treatments for injuries to neck, arms, and torso, where the text breaks off. Interestingly, ailments are approached in a purely scientific manner without inference to any supernatural forces except in one case that is mostly foreign to the papyrus. The papyrus also describes anatomical observations in exquisite detail. It contains the first known descriptions of the cranial sutures, the meninges, the external surface of the brain, the cerebrospinal fluid, and the intracranial pulsations. For example, the gyri and

1. A hieroglyph from the Edwin Smith papyrus translated as the membrane covering the brain.

sulci of the brain are beautifully described as "corrugations which form in molten copper... which the coppersmith pours off." The papyrus shows that the heart, blood vessels, liver, spleen, kidneys, ureters, and bladder were recognized, and that the blood vessels were known to be connected to the heart. Other vessels are described, some carrying air, some mucus, while two to the right ear are said to carry the breath of life, and two to the left ear the breath of death. However, the physiological functions of organs and vessels remained a complete mystery to the ancient Egyptians.

Aristotle of Stagira (384 B.C.E.-March 7, 322 B.C.E.) was an ancient Greek philosopher, student of Plato, and teacher of Alexander the Great. Charles Darwin regarded him as the most important previous contributor to the subject of biology. Aristotle's major contribution to biology was to attempt to understand natural phenomena in naturalistic terms, instead of believing them to be controlled by gods, demons, spirits, and supernatural forces, as was commonly believed in pre-scientific societies. Furthermore, he inaugurated a two-pronged scientific approach, that is, the application of disciplined thought to data acquired by observation. As Jonathan Barnes notes:

> Our modern notion of scientific method is thoroughly Aristotelian. Scientific empiricism – the idea that abstract argument must be subordinate to factual evidence, that theory is to be judged before the strict tribunal of observation – now seems a commonplace; but it was not always so, and it is largely due to Aristotle that we understand science to be an empirical pursuit. The point needs emphasizing, if only because Aristotle's most celebrated English critics, Francis Bacon and John Locke, were both staunch empiricists who thought that they were thereby breaking with the Aristotelian tradition. Aristotle was charged with preferring flimsy theories and sterile syllogisms to the solid, fertile facts. But the charge is outrageous; and it was brought by men who did not read Aristotle's own works with sufficient attention and who criticized him for the faults of his successors.

One of the most important achievements of Aristotle was to establish the field of comparative anatomy. The first three books of *Historia Animalium* (or *The History of Animals*), a treatise consisting of ten books, and the four books of *De Partibus Animalium* (or *On the Parts of Animals*), record the Aristotelian Anatomy. Aristotle could have obtained several of his descriptions only by frequently doing dissections himself, but he never performed any on the human body. Instead, he had to infer human anatomy from that of similar animals, "For the fact is that the inner parts of man are to a very great extent unknown, and the consequence is that we must have recourse to an examination of the inner parts of other animals whose nature in any way resembles that of man." The first anatomical drawing ever known was prepared and described by Aristotle. Although the actual illustration is lost, according to Aristotle it depicted the male urogenital system:

> All these descriptive particulars may be regarded by the light of the accompanying diagram; wherein the letter A marks the starting-point of the ducts that extend from the aorta; the letters KK mark the heads of the testicles and the ducts descending thereunto; the ducts extending from these along the testicles are marked MM; the ducts turning back, in which is the white fluid, are marked BB; the penis D; the bladder E; and the testicles XX.

2. Staff of Asclepius.

Alexandria in Egypt was the Greek center of learning where the first scientific studies designed to discover the workings of human anatomy were done, between 300 and 250 B.C.E. During this period, dissection of the human body was permitted by the state, shattering an age-old taboo. According to some authors, even human vivisection was practiced. It was said that two surgeons, Herophilus and Erasistratus, were allowed to dissect criminals who had been sentenced to death and who were delivered to them for this purpose by the state. Herophilus made many important anatomical observations, such as determining that the brain controls the nervous system, describing the duodenum, and discovering the lacteals and pineal gland. Erasistratus realized a number of important physiological principles, such as that arteries, veins, and nerves serve all parts of the body, that blood carries materials to those parts, and that the heart pumps materials around the body. Around 250 B.C.E., anatomical investigation of the human body once again became unthinkable and remained so until the late Middle Ages.

Claudius Galen (129 C.E.-ca. 216 C.E.) was not only a great physician but also a celebrated anatomist in the second century Greco-Roman world. Born in Pergamum in Asia Minor (today Bergama in western Turkey), Galen believed that Asclepius, the god of healing[i] in the ancient Greek pantheon, had saved him when he was ill with a near fatal condition due to an abscess, whereupon he became a servant of the god. He began studying medicine at the age of seventeen in Pergamum, finishing his studies at Smyrna, then at Corinth, and finally at Alexandria. Although human dissection was unacceptable in Galen's day, it was still possible to observe at least human skeletons in Alexandria. Galen's teachers, themselves trained in Alexandria, taught him the importance of anatomy for understanding the workings of the body.

In 157, Galen returned to Pergamum, where he was appointed physician to the local gladiators, a position that gave him the opportunity to practice surgical techniques. These skills were useful when in later life he conducted numerous dissections and vivisection experiments on animals, including apes, goats, dogs, pigs, and even an elephant. He used these investigations to infer human anatomy, some of which was wrong. Nevertheless, he made many important contributions to anatomy. For example, Galen showed by experiment that the arteries contain blood and not air as asserted five centuries earlier by Erasistratus. Also, by tying or cutting nerves at different locations along the spinal cord, he demonstrated their functions. Galen's findings were recorded in numerous medical treatises, made possible by his staff of shorthand writers who took down his words as he discoursed to friends, colleagues, and patients.

Galen looked upon science as a way of knowing and not as a corpus to be accepted on faith. Throughout his prodigious writings, he stressed the importance of accurate observation and experimental methodology:

> He [who wishes to know more than the multitude] must spend night and day learning thoroughly all that has been said by the most illustrious ancient authorities. Having learned all this, he must spend a prolonged period testing – observing what agrees with the ancient authorities and what does not – accepting one and rejecting the other.

Ironically, from the second to the seventeenth century, Galen's ideas became canonical, transformed into a mixture of medicine and philosophy called Galenism. Like Aristotle, Galen believed that the human body had been carefully designed by a provident and purposeful creator. For example, in a passage about the human hand, Galen wrote:

> Come now, let us investigate this very important part of man's body, examining it to determine not simply whether it is useful or whether it is suitable for an intelligent animal, but whether it is in every respect so constituted that it would not have been better had it been made differently…[T]he Creator gave us bone with the qualities it naturally possesses in order to support the fingers in each of the positions they assume…For if they were made without bones, they would do well only the work in which we need to curve them around the object to be grasped, and if they had just one bone, they would be of real use only in work where we need them extended. Since, however, they are not made

---

[i]   The staff of Asclepius entwined with a single serpent is today still an emblem of the healing art.

> without bones or with only one, but each finger has three of them articulating with on another, they therefore readily assume the positions necessary for all actions.

Galen's idea of an Intelligent Designer was compatible with both Muslim and Christian religious beliefs and thus promoted the preservation of his writings during later centuries, although Galen himself probably was pagan.[ii] After the fall of Rome (476), his texts were kept alive primarily by the Arabs until they were retranslated in Europe in the 1100s. But in translating his writings from Greek to Syriac, from Syriac to Arabic, from Arabic to Hebrew, and from Hebrew or Arabic to Latin, much of the spirit of Galen's work – especially his emphasis on observing for oneself rather than relying on authority – had been lost. Due to the extraordinarily expensive and tedious process of copying books, his voluminous works were abridged, repeating his anatomical conclusions but not his methods, empirical evidence, and doubts.

After the demise of Roman civilization, the Christian Church soon became the dominant religious, political, and intellectual force. During the Middle Ages, the Church considered the study of nature, including the human body, to be a waste of time compared to the study of the Bible. As the Church father Tertullian (ca. 160-225) wrote, "Investigation since the Gospel is no longer necessary." Knowledge other than that which made a man "wise unto salvation" was useless, since the apocalyptic Day of Judgment was believed to be imminent. St. Augustine (354-430) and most other theologians denounced dissection of the human body. Because the Bible taught that man was made in the image of God, the act of dissecting a corpse was considered to be tantamount to dissecting God. In fact, there is no question that Augustine discouraged free inquiry about earthly matters, as evidenced by his writings:

> There is another form of temptation, even more fraught with danger. This is the disease of curiosity…It is this which drives us on to try and discover the secrets of nature, those secrets which are beyond our understanding, which can avail us nothing and which man should not wish to learn…In this immense forest, full of pitfalls and perils, I have drawn myself back, and pulled myself away from these thorns. In the midst of all these things which float unceasingly around me in everyday life, I am never surprised at any of them, and never captivated by my genuine desire to study them…I no longer dream of the stars.

As a result, no advances in the study of human anatomy occurred during the Middle Ages. Instead, the Church declared Galen to be the infallible authority on the subject, a position that he arguably would have refused.

In the twelfth and thirteenth centuries, the first European universities were established in Italy, then France, and later Britain and the Germanic countries. Probably the birth of modern anatomy can be traced to the University of Bologna in northern Italy, which began as a law school but had a faculty of medicine as early as 1156. The first human dissections here, probably in the middle of the thirteenth century, were post-mortem examinations for legal purposes. The first known autopsy was recorded by a surgeon and teacher at Bologna, William of Saliceto, in his book *Surgery and Anatomy*. He

---

[ii]   Galen believed that the Intelligent Designer has not created matter, but arranges matter in the best possible way. He argued against the Judeo-Christian belief in an omnipotent god who, by his mere command, could create the world *de novo*.

mentioned a post-mortem conducted in about 1275 to determine whether certain chest wounds were the cause of death. Other similar forensic autopsies were mentioned in 1295 and 1300. No doubt, many more were performed and recorded in documents that have not survived. Before long, these medico-legal investigations developed into dissections for anatomical purposes.

The earliest evidence for the dissection of human cadavers for purely anatomical purposes was provided by Mundinus, also known as Mondino de Luzzi, a professor at the University of Bologna. In a manual on human anatomy entitled *Anathomia* (or *De Anatome*), he wrote, "A woman I anatomized last year, that is in the year of Christ, 1315, in the month of January, had a womb double as big as her that I anatomized in March of the same year." However, the booklet by Mundinus did not depart from Galenic anatomy. By 1341, the practice of public dissection in their medical faculties had spread from Bologna to Padua and then to Pavia, Florence, Siena, and other Italian schools.

Nevertheless, for about two hundred years, apparently no one noticed the mistakes Galen had made in human anatomy by relying on animal dissections. A tradition had emerged in which a professor would mount his podium and sit high on his ornate chair, like a bishop on his throne. (This is the origin of the "chair" of a university, the actual seat of the eminent professor.) The professor would read from the grossly inaccurate works of the ancients or the more recent rewrites of Galen, such as the *Anathomia* by Mundinus, while an assistant, called the ostensor, stood below. The ostensor never actually touched the corpse himself but, with a baton, pointed out structures as they were mentioned by the professor. The dissection was done by a demonstrator, who was either a menial servant or perhaps a barber or surgeon, both of whom rated quite low in the medical hierarchy of the time. Often the ostensor could not find the organ as described, but invariably the body rather than Galen was held to be in error. There was no point in the professor looking for himself at the cadaver, since everything worth learning could be found in Galen's books. The students stood around the corpse, watching and listening, but mainly listening. The fifteenth-century illustration on the back cover of this text shows the professor in his chair, reading from a book, while a menial demonstrator dissects as directed by the wand of the ostensor. The students in academic dress take no active part in the dissection.

Leonardo da Vinci (1452-1519) contributed to anatomy mainly by creating a system of drawing that enabled anatomists, and even modern-day physicians, to transmit their findings to students. His anatomical career began while he was apprenticed to Andrea del Verrocchio in Florence, who insisted that all his pupils learn anatomy. In 1495, Leonardo abandoned his anatomical investigations, believing they had fulfilled his artistic needs. In the winter of 1507-1508, his interest in anatomical investigation was revived by chance during one of his visits to the hospital Santa Maria Nuova in Florence, where he witnessed the death of an old man. Of this event, Leonardo wrote:

> ...an old man a few hours before his death told me that he had passed a hundred years, and that he did not feel any bodily deficiency other than weakness. And thus while sitting on a bed in the hospital of Santa Maria Nuova in Florence, without any movement or sign of distress he passed away from his life. And I made an anatomy of him in order

> to see the cause of so sweet a death. This anatomy I described very diligently and with great ease because of the absence of fat and humors which much impede knowledge of the parts.

Leonardo continued his dissections until, in 1515, he was accused by a German mirror maker named Giovanni degli Specchi of sacrilegious practices. As a result, Leonardo was banned from doing anatomical investigations by Pope Leo X, just four years before his death. Nevertheless, during his lifetime, he had dissected over thirty human corpses of different ages, both male and female, producing some 750 drawings, including studies of bone structures, muscles, internal organs, the brain and even the intrauterine position of the fetus. His studies of the heart suggest that he was on the verge of discovering the concept of the circulation of the blood. As Osler observes, "Leonardo was the first of modern anatomists, and fifty years later, into the breach he made, Vesalius entered."

On December 6, 1537, Andreas Vesalius (1514-1564) was elected by the academic senate to the chair of surgery and anatomy at the University of Padua. He broke with tradition and began dissecting human corpses for himself to show his students the fine details of anatomy. In the process, he discovered that Galenic anatomy was based on the animal body and consequently, much of what Galen had presented as human anatomy was wrong. Vesalius then courageously set out to put together a new anatomy book that included his discoveries. Published in 1543, *De humani corporis fabrica libri septem* (or *The Seven Books on the Structure of the Human Body*) included detailed engravings and descriptions that corrected a thousand years of ignorance. Ironically, this achievement was due precisely to the rediscovery of Galen's own principle: question authority and trust only your own observations.

REFERENCES.

Alvey, R. Kevin. "The anatomical drawings of Leonardo da Vinci." Feb. 19, 2006 <www.geocities.com/CollegePark/1070/leonardo.html>.

Aristotle. *Historia Animalium*, trans. D'Arcy Wentworth Thompson, Book I: Part 16; Book III: Part 1.

Barnes, Jonathan. *Aristotle: A Very Short Introduction* (Oxford, UK: Oxford University Press, 2000).

Breasted, James H. *The Edwin Smith surgical papyrus.* (Chicago: University of Chicago Press, 1980).

Grmek, Mirko D., ed. *Western Medical Thought from Antiquity to the Middle Ages* (Cambridge, MA: Harvard University Press, 1998).

Freeman, Charles. *The Closing of the Western Mind* (New York, NY: Vintage Books, 2002).

Galen. *My Own Books* (Kühn vol. xix, p. 19)

—. *On the Usefulness of the Parts of the Body*, trans. Margaret Tallmadge May (Ithaca, NY: Cornell University Press, 1968).

Knight, Bernard. *Discovering the Human Body* (New York, NY: Lippincott and Crowell, Publishers, 1980).

Moore, John A. *Science as a Way of Knowing* (Cambridge, MA: Harvard University Press, 1993).

Nutton, Vivian. "Logic, learning, and experimental medicine." *Science* 295: 800-1 (2002).

Osler, William. *The Evolution of Modern Medicine.* (1921).

St. Augustine. *Confessions.*

Temkin, Owsei. *Galenism: Rise and Decline of a Medical Philosophy* (Ithaca, NY: Cornell University Press, 1973).

Tertullian. *The Prescription against Heretics*, chapter VII, "Pagan Philosophy the Parent of Heresies."

TOPIC 1. THE NATURE OF SCIENCE

The ancient Greeks are often given credit for developing the skeptical, inquiring, experimental method of science. For example, they apparently first proposed the idea that laws of Nature, rather than capricious gods, govern the world. Titus Lucretius (about 99-55 B.C.E.) summarized their views, "Nature free at once and rid of her haughty lords is seen to do all things spontaneously of herself without the meddling of the gods." But that does not mean that science would never have emerged without the ancient Greeks. The astronomer Carl Sagan (1934-1996) pointed out that even so-called primitive peoples, such as the !Kung San of the Kalahari Desert in Africa, have demonstrated aspects of scientific thinking and concluded, "A proclivity for science is embedded deeply within [humans], in all times, places and cultures."[1]

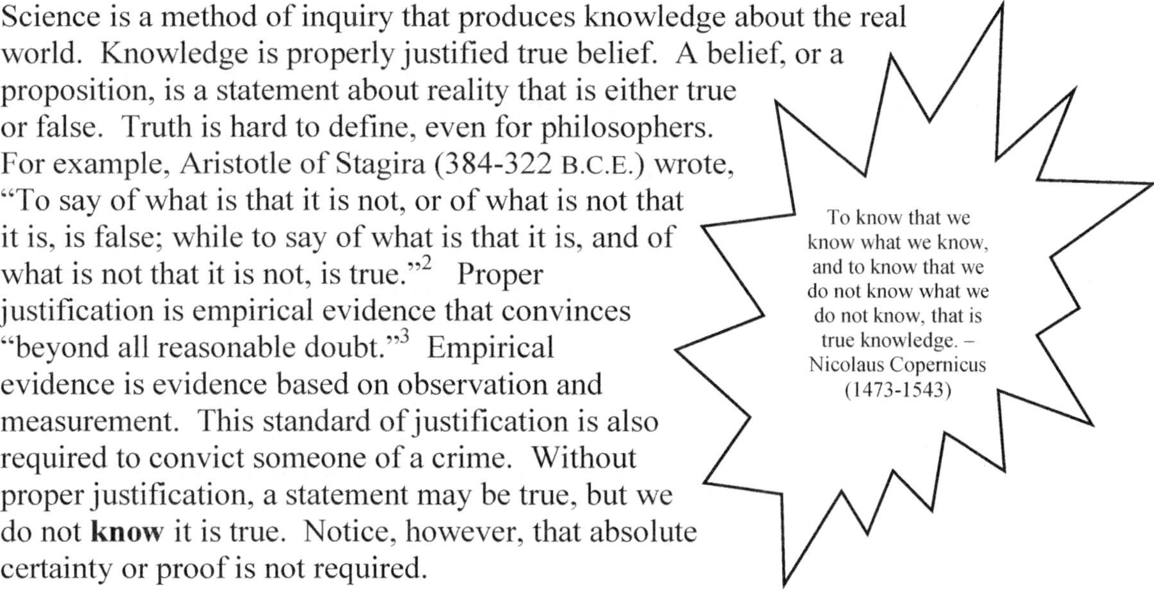

Science is a method of inquiry that produces knowledge about the real world. Knowledge is properly justified true belief. A belief, or a proposition, is a statement about reality that is either true or false. Truth is hard to define, even for philosophers. For example, Aristotle of Stagira (384-322 B.C.E.) wrote, "To say of what is that it is not, or of what is not that it is, is false; while to say of what is that it is, and of what is not that it is not, is true."[2] Proper justification is empirical evidence that convinces "beyond all reasonable doubt."[3] Empirical evidence is evidence based on observation and measurement. This standard of justification is also required to convict someone of a crime. Without proper justification, a statement may be true, but we do not **know** it is true. Notice, however, that absolute certainty or proof is not required.

A scientific statement is true if it has been tested extensively and elegantly and has not been falsified, but it is never more than "the most accurate statement that can be made with the evidence at hand."[4] New evidence may change previous conclusions. Therefore, scientific knowledge (as well as jury verdicts) is said, philosophically speaking, to be provisional.[5] For example, the phlogiston theory was the best explanation of combustion until September 5, 1775. On that day, Antoine-Laurent Lavoisier (1743-1794) presented his oxygen theory of combustion to the French Academy of Science in a paper entitled *Memoir on Combustion in General*. Lavoisier provided evidence from quantitative experiments in which he had used a highly accurate balance, which had not been done by his predecessors. (Likewise, sometimes convicted persons are shown to be innocent when new evidence is later discovered.)

Nevertheless, some of our fundamental knowledge today is as certain as is our knowledge of tables and chairs. For example, for thousands of years, people believed that the Sun revolves around the Earth, an idea known as the Theory of Geocentrism, because it seemed intuitively true: our senses tell us so. However, today we know beyond all reasonable doubt that the Earth revolves around the Sun, an idea known as the Theory of Heliocentrism. There is virtually no chance that the heliocentric theory will ever be falsified; certainly, scientists will never return to a geocentric theory. Likewise, there is virtually no chance that the existence of gravity, atoms, genes, and evolution will ever be falsified because the evidence in favor of each of these things is so overwhelming.

Certain beliefs, or assumptions, are accepted in science without evidence because we could not even begin to do science (or even get out of bed in the morning!) without them.[6] These assumptions include:

This cosmos, the same for all, was made by neither god nor man, but was, is and always will be: an everlasting fire, kindling and extinguishing according to measure. – Heraclitus of Ephesus (*fl.* 500 B.C.E.)

- There is an external world separable from our perception; objective reality exists. There really are things out there. Everything is not simply a figment of the imagination.
- The operation of the universe is regular and predictable. If phenomena occurred at random, without any warning or pattern, no amount of analysis would uncover any regularity to them.
- The basic patterns that describe events in the natural world are the same throughout the universe and throughout time (the Cosmological Principle).
- The universe is knowable; no aspects of the universe are beyond human understanding. As the physicist Albert Einstein said, "The most incomprehensible thing about the world is that it is comprehensible."[7]
- Scientific knowledge, in principle, can be confirmed by anyone.

Some scientific disciplines have more specific assumptions:

- Geologists assume that the history of the Earth is best interpreted in terms of what is known about geological processes at work in the present, rather than supposed processes in the past (the Principle of Uniformitarianism).
- Biologists assume that all forms of life reproduce, grow, and respond to changes in the environment.

Some forms of propositions in science include the law, the hypothesis, the theory, and the fact. A law is "a descriptive generalization about how some aspect of the natural world behaves under stated circumstances."[8] In other words, it is a description of some regularity in nature that is universal and without exception. Laws are quite rare in science and are found mainly in physics and chemistry. They can often be expressed in mathematical form, such as the first law of thermodynamics: $\Delta E = q + w$.

A hypothesis is "a tentative statement about the natural world leading to deductions that can be tested."[9] It is a possible explanation of some set of facts, giving us at least initial understanding of what we observe. Scientific explanations must be falsifiable to be scientific. For example, the idea that gods or invisible demons cause epilepsy cannot be falsified because gods could just as well conceal their involvement from human detection, and invisible demons would be undetectable. On the other hand, the idea that abnormal neurological activity causes epilepsy can be falsified because brain waves can be recorded during a seizure and the abnormal activity observed. The rule that only natural causes can be invoked in scientific explanations, because supernatural causes are not empirical, is called the Principle of Methodological Naturalism. A Greek physician living about the time of Hippocrates applied this rule to the cause of epilepsy: "It seems to me that the disease called sacred has a cause, just as other diseases have. Men think it divine merely because they do not understand it. But if they called everything divine that they did not understand, there would be no end of divine things! …In Nature all things are alike in this, in that they can be traced to preceding causes."[10] The requirement for falsifiability and empirical causes is why Darwin said, "On the ordinary view of the independent creation of each being, we can only say that so it is; – that it has pleased the Creator to construct all the animals and plants in each great class on a uniform plan; but this is not a scientific explanation."[11]

Hypotheses whose predictions have been tested often and have never failed are sometimes combined into general statements called theories. A theory, such as the theory of gravity and atomic theory, is a well-substantiated, unifying explanation for a broad range of observations. The theory of evolution explains how life on Earth has changed and is one of these well-supported explanations. Theories are the end-points of science – that of which we are most certain. Thus, scientists do not use the word theory as does the general public, to whom the word implies a **lack** of knowledge or a guess. Finally, a theory cannot become a law (explanations don't become descriptions).

A fact is an observation that has been repeatedly confirmed and is accepted as true beyond a reasonable doubt. Scientists also use the word in the sense of "something that has been tested or observed so many times that there is no longer a compelling reason to keep testing or looking for examples."[12] In this sense, evolution is also a fact. As the Society for the Study of Evolution states, "'Evolution' refers both to a set of scientific facts and to a theory explaining such facts. 'Evolution' refers to the scientific fact that biological organisms have changed through time, and that all life, including humanity, has descended with modification from common ancestors. Evolution is as well documented as are other currently accepted scientific facts. The theory of evolution is a comprehensive and well-established scientific explanation, based on natural processes, of the fact of biological evolution."[13] Stephen Jay Gould wrote, "Well, evolution is a theory. It is also a fact. And facts and theories are different things, not rungs in a hierarchy of increasing certainty. Facts are the world's data. Theories are structures of ideas that explain and interpret facts. Facts don't go away when scientists debate rival theories to explain them. Einstein's theory of gravitation replaced Newton's in this century, but apples didn't suspend themselves in midair, pending the outcome. And humans evolved from ape-like ancestors whether they did so by Darwin's proposed mechanism or by some other yet to be discovered."[14]

At the heart of science is an essential balance between two modes of thought: creative thinking and skeptical, or critical, thinking.[15] Creative thinking is defined as the ability to see relationships between phenomena or ideas that were not previously obvious. As the physicist Erwin Schrödinger (1887-1961) said, "Thus, the task is, not so much to see what no one has yet seen; but to think what nobody has yet thought, about that which everybody sees."[16] The American philosopher John Dewey expressed it this way: "Every great advance in science has issued from a new audacity of imagination. What are now working conceptions, employed as a matter of course because they have withstood the tests of experiment and have emerged triumphant, were once speculative hypotheses."[17]

Critical thinking is defined as the "means to construct, and to understand, a reasoned argument and – especially important – to recognize a fallacious or fraudulent argument. The question is not whether we **like** the conclusion that emerges out of a train of reasoning, but whether the conclusion **follows** [logically] from the premise or starting point and whether that premise is true."[18] In 1605, the English philosopher Francis Bacon (1561-1626) described critical thinking as a "desire to seek, patience to doubt, fondness to meditate, slowness to assert, readiness to consider, carefulness to dispose and set in order; and hatred for every kind of imposture." René Descartes (1596-1650), the father of modern philosophy, wrote, "I did not imitate the skeptics who doubt only for doubting's sake, and pretend to be always undecided; on the contrary, my whole intention was to arrive at a certainty, and to dig away the drift and the sand until I reached the rock or the clay beneath."[19] David Hume (1711-1776) wrote, "No testimony is sufficient to establish a miracle, unless the testimony be of such a kind, that its falsehood would be more miraculous, than the fact, which it endeavors to establish; and even in that case there is a mutual destruction of arguments, and the superior only gives us an assurance suitable to that degree of force, which remains, after deducting the inferior."[20] Carl Sagan wrote, "Extraordinary claims require extraordinary evidence."[21]

Skepticism in science is institutionalized as peer review. When a research paper is submitted to a reputable scientific journal for publication, the editor sends it to several peer reviewers. Peer reviewers are other scientists who, based on their knowledge of the subject matter, are competent to judge the quality of the research. The role of the reviewer is to ensure that the experimental design was appropriate, that the conclusions are justified, and that the research is useful enough to science to merit publication of the paper. The editor uses the advice of all the reviewers to decide if the paper will be published.

Science is based on a logical approach to explaining nature. In logic, an argument consists of a claim (conclusion) and one or more reasons (premises) given to support that claim. Two forms of reasoning are used: inductive and deductive.

Inductive reasoning involves an argument in which the logical connection between premises and conclusion is one of probability. That is, if the premises of an inductive argument are true, the conclusion is **likely**, but not necessarily, true. For example: "All swans that have been observed are white. Therefore, all swans are white." This conclusion was accepted as true by people in the northern hemisphere until black swans were discovered in Australia. Another example: "The sun has risen every day in recorded history. Therefore, the sun will rise tomorrow." Bertrand Russell (1875-1970) wrote about induction, "We know that all these rather crude expectations of uniformity are liable to be misleading. The man who has fed the chicken every day throughout its life at last wrings its neck instead, showing that more refined views as to the uniformity of nature would have been useful to the chicken."[22] Nevertheless, scientists assume that there are general rules that apply to natural phenomena and that often one needs but a small sample to find rules with broad applicability.[23]

Deductive reasoning involves an argument in which the logical connection between premises and conclusion is one of necessity. That is, if the premises of a deductive argument are true, the conclusion **must** be true. For example: "All men are mortal. Socrates is a man. Therefore, Socrates is mortal." Another example: "All poodles are dogs. All dogs are animals. Therefore, all poodles are animals." If the premises are inadequate to support the conclusion, then a logical fallacy has been committed.[24, 25, 26]

The scientific method is a procedure that allows scientists to systematically approach a problem or interesting observation. It is often described by anywhere from four to seven "steps," depending on the author. I will break the scientific method down into the following five steps:

1) Ask a question or identify a problem to be solved, based on observations of the world (or very specific parts of the world).
2) Propose multiple hypotheses that are consistent with the observations. Scientists follow the Principle of Parsimony. This is the idea that whenever two explanations are equally consistent with the facts, the simpler explanation is more likely to be correct. The principle can also be expressed as "causes should not be multiplied beyond necessity." Since the simpler hypothesis is also usually easier to test, this is the one the scientist begins with.
3) Deduce predictions that must also be true if a given hypothesis is true. That is, predict other necessary consequences besides the known evidence upon which the hypothesis was originally founded. If such additional lines of evidence are confirmed by subsequent tests, then the hypothesis may be considered increasingly probable.
4) Test those predictions by experiments or further observations. An experiment is a set of actions and observations, performed in the context of solving a particular problem or question, to support or falsify a hypothesis concerning phenomena. An experiment may be conducted under natural or artificial conditions.
5) Analyze the results of the test(s) and draw conclusions. If the results support the hypothesis, deduce another prediction from the hypothesis and test it. If the results refute the hypothesis, test an alternative hypothesis. As Einstein said, "No amount of experimentation can ever prove me right; a single experiment can prove me wrong."[27] (However, scientists in practice do not **immediately** reject a promising hypothesis after a few negative experimental results; they have to decide if the problem lies with the hypothesis or with the experiment(s).[28, 29] For example, the English physicist and chemist Robert Boyle (1627-1691) once had to perform an experiment 50 times before he finally got positive results.[30])

Thus, most scientific research actually involves eliminating false hypotheses. As Sir Arthur Conan Doyle's famous detective Sherlock Holmes remarked, "When you have eliminated the impossible, whatever remains, *however improbable*, must be the truth."[31] For example, Matthew Meselson and Franklin Stahl performed an elegant experiment in 1958 that simultaneously tested three alternative hypotheses that explain the way DNA replicates; the results were inconsistent with two of the hypotheses.[32]

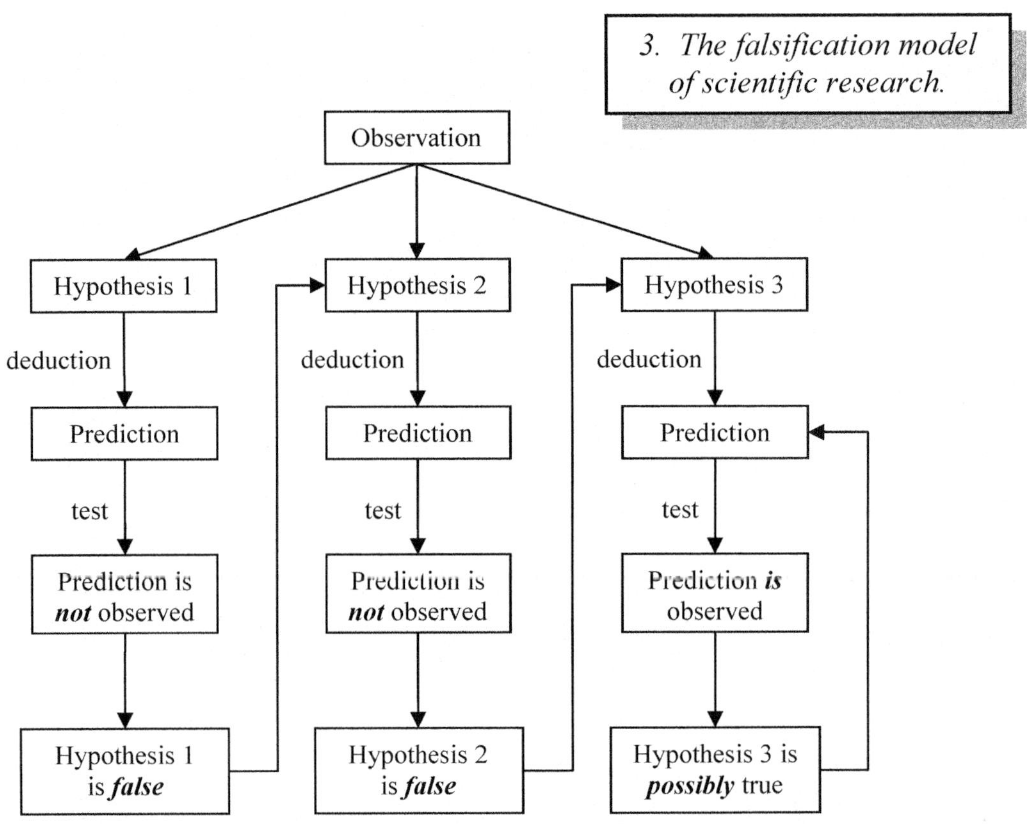

The simplest form of experiment tests the hypothesis that a single preceding event is the cause of a single subsequent observation. To make sure a test of a hypothesis is objective, significant variables must be isolated and the experiment must be repeatable.

The phenomenon that is being studied in a scientific experiment is called the **dependent variable**, while the factors that potentially influence it are called **independent variables**. A condition that remains the same throughout the experiment is called a **controlled variable**. Thus, a hypothesis proposes that an independent variable causes the dependent variable. To borrow a simple example from physics, suppose we want to test the hypothesis that the time it will take for a ball to drop to the bottom of a container filled with fluid is directly related to the density of that fluid. The density of the fluid is the independent variable, while the time of drop is the dependent variable, since the density determines the time. The force of gravity, although important, does not change and is therefore a constant in the experiment. (However, note that correlation does **not** imply causality; in other words, a relationship between two variables should not be accepted as cause-and-effect unless a well-founded explanation can be offered for the causal relationship. For example, people who wear skirts get breast cancer more often than people who never wear skirts, but this does not mean that wearing skirts causes breast cancer.)

Most research involves more complicated situations in which many variables are involved and isolating the effect of just one is difficult. A solution to this problem is to divide the study into a **control group** of subjects and an **experimental group** of subjects. Both groups are treated the same way, with the exception of one independent variable. Conditions other than the independent and dependent variables are not allowed to change throughout the experiment. Thus, the control group serves as a benchmark that allows the scientist to decide whether the predicted effect is really due to the independent variable. This is called a **controlled experiment**. In some experiments, it is a good idea to have both a **positive control** and a **negative control**. The positive control confirms that the basic conditions of the experiment were able to produce a positive result, even if none of the actual experimental samples produce a positive result. The negative control demonstrates the base-line result obtained when a sample does not produce a measurable positive result.

A famous early controlled experiment in biology was performed by the Italian physician Francesco Redi (1621- 1697) to test his hypothesis that flies lay eggs that hatch into maggots in spoiled meat. Before Redi, the appearance of maggots in spoiled meat was considered evidence of spontaneous generation, the production of life in complex form from nonliving matter over a matter of days. Many different conditions might influence the appearance of maggots, including not only the presence of flies, but also temperature, time of year, and whether the experiment was conducted indoors or outdoors. Redi took two clean jars and filled them with similar pieces of meat. He left one jar open (the control sample) and covered the other with gauze to keep out flies (the experimental sample). He did his best to keep all the other variables the same for both jars. Redi observed after a few days that maggots swarmed over the meat in the open jar, but no maggots appeared in the covered jar, although he reported that the meat nevertheless became "putrid and stinking" (due to bacteria). Redi concluded that his hypothesis was correct, and that meat did not spontaneously generate maggots.

As mentioned previously, scientists also insist on the ability to reproduce results. The investigator conducting the experiment may in some way, consciously or unconsciously, influence its outcome. A carefully designed and described experiment enables other scientists to repeat it and thus confirm or challenge its findings.

The standard experiment to determine the effectiveness of a medical treatment is called the "randomized controlled trial" (RCT). Subjects who will be part of the experiment are randomly selected from the population. The population consists of all individuals who possess characteristics, called inclusion criteria, which will **not** favorably or unfavorably affect the outcome of the experiment. Subjects are assigned randomly to the control or treatment groups. Such random selection reduces bias in research by eliminating as much as possible the subjective decisions made by the scientist. Ideally, neither scientist nor subjects should know who is being treated or not treated until the experiment is completed. This is called "double-blind" testing.

Table 1. Design of Experiment done by Francesco Redi

	Control Sample	Experimental Sample
Controlled Variables (What is kept the same among all samples.)	• Clean jars • Meat • Kept outdoors under identical conditions	
Independent Variable (What is deliberately changed among samples.)	• Jar left open	• Jar covered with gauze to keep out flies
Dependent Variable(s) (What is observed as different among samples.)	• Maggots appear	• No maggots (although meat became "putrid and stinking")
Uncontrolled Variables (What is accidentally or unintentionally different among samples.)	• Scientist error • Variation in sample preparation • ?	• Scientist error • Variation in sample preparation • ?

TOPIC 2. WHAT ARE HUMANS?

For centuries, the theological notion that "man" was literally made in the "image of God" meant that he could not be an animal, because God is not an animal. Within the context of the Christian view, man saw himself as so unique that it did not occur to him to inquire into the possibility of a relationship with the animals in any derivative sense; he merely shared God's world with them.

Despite being a self-avowed creationist, Carolus Linnaeus (1707-1778), often called the father of modern taxonomy, recognized that **man is an animal**: he named our species *Homo sapiens*. Setting the stage for Darwin, Linnaeus also placed humans in the order Primates (a larger, more inclusive category than our genus) along with all of the apes, monkeys, and prosimians. He got into hot water for this classification. In response to criticism, Linnaeus wrote to a colleague, Johann Gmelin, in 1747, "I ask you and the whole world for a generic difference between man and ape which conforms to the principles of natural history. I certainly know of none...If I were to call man ape or vice versa, I should bring down all the theologians on my head. But perhaps I should still do it according to the rules of science."[iii] The fact that humans are animals is important

---

[iii] However, Linnaeus did **not** say that man is **just** an ape. As the twentieth-century paleontologist George Gaylord Simpson noted, "These fallacies arise from what Julian Huxley calls 'the-nothing-but' school. It was felt or said that because man is an animal, a primate, and so on, he is *nothing but* an animal or *nothing but* an ape with a few extra tricks. It is a fact that man is an animal, but it is not a

medically. It means, for example, that insulin extracted from the pancreas of pigs could once have been used to treat human diabetes. It also means that new treatments for human diseases can be tested on certain animals first for safety and effectiveness.

Following Linnaeus' example, biologists who classify living and extinct organisms have included humans within a hierarchy[33], shown in Table 2, in which each level is a smaller subgroup of the one above it and the members of each group share certain characteristics. The order Primates also contains the apes, monkeys, lemurs, and a few other species. The superfamily Hominoidea (the hominoids) consists of the apes and humans and includes the family Hominidae (the hominids), which consists of the **great** apes and humans. The hominids, in turn, are divided into three subfamilies. The subfamily Ponginae includes only one species, the orangutan (*Pongo pygmaeus*). The subfamily Gorillinae also includes just one species, the gorilla (*Gorilla gorilla*), which is the largest living primate. The subfamily Homininae consists of the chimpanzees and humans and is divided into two tribes. The tribe Panini contains the two species of chimpanzee: the common chimpanzee (*Pan troglodytes*) and the pygmy chimpanzee, or bonobo (*Pan paniscus*). The tribe Hominini (the hominins) consists of the human and human-like hominids.

Table 2. The Classification of Humans

Level	Group	Shared characteristics
Kingdom	Animalia	Eukaryotic, multicellular, chemoheterotrophic, and embryonic development
Phylum	Chordata	A notochord, dorsal hollow nerve cord, pharyngeal slits, and a postanal tail
Subphylum	Vertebrata	A jointed vertebral column, a cranium , and a well-developed brain and sense organs
Class	Mammalia	Mammary glands, hair, and three middle ear bones
Order	Primates	Shortened rostrum and forwardly directed orbits, associated with stereoscopic vision; opposable hallux and pollex; unfused and highly mobile radius and ulna in the forelimb and tibia and fibula in the hindlimb; pentadactyl feet and presence of a clavicle
Family	Hominidae	The largest primates, with robust bodies and well-developed forearms, flattened nails on all digits, and a large braincase; the dental formula is 2.1.2.3/2.1.2.3 = 32
Genus	*Homo*	Large brain, small teeth, and tool-making
Species	*Homo sapiens*	Very large brain, lighter build, distinctive cranium

fact that he is nothing but an animal" [emphasis in the original]. G. G. Simpson, *The Meaning of Evolution* (New Haven: Yale University Press, 1952) 233.

The tribe Hominini includes both living and extinct species belonging to several genera, including the extinct genus *Australopithecus* and the genus *Homo*. Australopithecines had limbs with adaptations for bipedal locomotion (habitual walking on two legs, also called bipedalism) and had reduced canines, but they did not have the large brain size or characteristic pelvis and femur shape of the *Homo* species. Several fossil species are known including †*A. afarensis*[iv], which lived from 4.0 to 3.0 million years ago (Mya), and †*A. africanus*, which lived from 3.3 to 2.5 Mya. Members of the genus *Homo* have relatively large brains and small molars, canines, and limb bones compared to other apes. Several fossil species and one extant species are known including: †*H. habilis*, which lived from 2.5 to 1.9 Mya; †*H. rudolfensis*, which was contemporary with *H. habilis*; †*H. ergaster*, which lived from 1.9 to 1.6 Mya; †*H. erectus*, which lived from 1.6 million to 50,000 years ago; †*H. heidelbergensis*, which lived from 800,000 until the appearance of modern humans; †*H. neanderthalensis*, which lived from 200,000 to about 30,000 years ago; and *H. sapiens*, the only living species, which appeared over 130,000 years ago.[34]

TOPIC 3. BIOLOGICAL EVOLUTION

For many years before Darwin, anatomists had noticed that the body structures of different animal species are often similar in one of two different ways: homology or analogy.

Homology (what naturalists used to call "true affinity"[35]) is "the relationship arising from identity of structure without reference to function."[36] Homologous structures are the morphologically similar structures that may or may not be functionally similar in different species. For example, as illustrated on the frontispiece of this book, the French naturalist Pierre Belon in the sixteenth century published a treatise on birds in which he showed, side by side, the skeletons of a bird and of a human being. He gave the same names to the bones of the bird that had human counterparts. There was no difficulty in demonstrating the correspondence of bones in two such apparently different structures, for example, the bird's wing and the human arm. The proximal bones of the human arm and of the bird wing are modifications of the "same thing."[37] Likewise, the forelimbs of frogs and alligators contain the same set of bones modified for crawling/swimming. The malleus and incus bones, which are two of the three auditory ossicles in the mammalian middle ear, and the much larger bones of the reptilian jaw are another example of homologous structures (see Topic 24.7).[38, 39]

Analogy is the relationship arising from "similarity of function, without reference to structure."[40] Analogous structures are functionally similar but morphologically different in different species. For example, the wings of birds and insects have totally different components, yet they serve the same function and have a similar shape. The sleek, streamlined, fat-insulated shapes of seals and penguins are another example of analogous structures.

However, for centuries these similarities had been interpreted as evidence of the "thoughts" of a supernatural being, or Intelligent Designer. Louis Agassiz was perhaps

---

[iv]   The symbol "†" indicates that the species is extinct.

the last reputable scientist who advocated the idea of divine intervention in natural history. In 1857, he wrote, "[A]s long as it cannot be shown that matter or physical forces do actually reason, I shall consider any manifestation of thought as evidence of the existence of a thinking being as the author of such thought, and shall look upon an intelligent and intelligible connection between the facts of nature as direct proof of the existence of a thinking God, as certainly as man exhibits the power of thinking when he recognizes their natural relations."[41]

Although others had considered biological evolution before them, the English naturalists Charles Darwin and Alfred Russell Wallace independently were the first to suggest a reasonable explanation for it. Their papers were simultaneously presented on July 1, 1858 at a meeting of the Linnaean Society of London. Darwin in 1859 published "one long argument" for evolution through natural selection entitled, *The Origin of Species*. He brought together such overwhelming evidence that within ten years after 1859 hardly a competent biologist was left who did not accept the fact of evolution.[42] However, his explanation for evolution, natural selection, was rejected by most biologists until they began to understand the genetic basis for heredity. Population genetics, which emerged in the 1930s and 40s, was combined with Darwin's theory of evolution to provide a powerful explanation of biological evolution, now called the "modern synthesis."[43] In 1871, Darwin published *The Descent of Man*, which applied his theory of evolution to human beings.

Evolution today is defined as the change over time in the genetic makeup, or **gene pool**, of a population of organisms or viruses. Examples include the evolution of antibiotic-resistant bacteria, of new strains of HIV, of insects resistant to insecticides, and of new species of organisms. A heritable trait that improves the ability of an individual to survive and reproduce in its environment is called an **adaptation** or adaptive trait. However, if the environment changes, a given trait may no longer be adaptive. An adaptation can be an anatomical structure, physiological process, or behavior. Examples of adaptations in animals are camouflage, disease resistance, and sharp claws or fangs. An adaptive trait is heritable because it is determined by one or more genes that are passed down from one generation to the next. Genes are short segments of DNA (see Topic 6) that determine the observed traits in an individual. All the genes in all the individuals in a population are collectively called its gene pool. A permanent change in the DNA of a gene is called a **mutation**. Most inherited mutations are harmful or, at best, harmless to the individual. However, the occasionally useful mutation is the ultimate source of variation in a population that can lead to the formation of new adaptations with the passage of time.

Individuals with an adaptive trait generally produce more offspring than those individuals without it, a process called natural selection. The offspring inherit the gene(s) for the adaptive trait, bringing about an increase in their frequency in the gene pool of the next generation that defines evolution. Natural selection is the only evolutionary process that preserves adaptations. The conditions in the environment that cause individuals with an adaptive trait to be more reproductively successful than individuals without it are called **selective pressures**. Examples include climate, predators, pathogens, and limited resources.

Complex adaptive traits, such as the vertebrate eye, can evolve because natural selection occurs gradually, over geologic time. Darwin observed that, "With animals such as the giraffe, of which the whole structure is admirably co-ordinated for certain purposes, it has been supposed that all the parts must have been simultaneously modified; and it has been argued that, on the principle of natural selection, this is scarcely possible. But in thus arguing, it has been tacitly assumed that the variations must have been abrupt and great. No doubt, if the neck of a ruminant were suddenly to become greatly elongated, the fore limbs and back would have to be simultaneously strengthened and modified; but it cannot be denied that an animal might have its neck, or head, or tongue, or fore-limbs elongated a very little without any corresponding modification in other parts of the body; and animals thus slightly modified would, during a dearth, have a slight advantage, and be enabled to browse on higher twigs, and thus survive. A few mouthfuls more or less every day would make all the difference between life and death. By the repetition of the same process, and by the occasional intercrossing of the survivors, there would be some progress, slow and fluctuating though it would be, towards the admirably coordinated structure of the giraffe."[44] The gradual modification of an ancestral structure into a new one through a series of steps, each of which is an adaptation, is called **cumulative natural selection** (see Figure 49).

When the gene pools of two populations of the (initially) same species accumulate enough genetic differences such that the populations can no longer interbreed and produce viable, fertile offspring, they are said to be separate species. At least one new species is produced when this occurs, so this is called branching evolution, or **cladogenesis**. The number of living species increases as a result of branching evolution. Branching evolution can explain the existence of homologous structures. They are produced when identical anatomical structures in both species, inherited from their common ancestor, are modified for different functions due to different selective pressures. This process is called **parallel evolution**. On the other hand, analogous structures are produced when, after divergence from the common ancestor, different anatomical structures are modified for the same function in both species due to similar selective pressures. This process is called **convergent evolution**.

TOPIC 4. THE EVIDENCE FOR EVOLUTION

The evidence establishes beyond a reasonable doubt that living things, including humans, have evolved.[45] One source of evidence is the **adaptive compromise**, a modification of an ancestral structure to form a new one with less than optimal results. Adaptive compromises are extremely common because they are the easiest way to get something new without interfering with functions that must continue in the organism. To use an analogy, suppose you live in a two-bedroom house and your family grows so that you need a four-bedroom house. The easiest way to get the additional bedrooms is to remodel, rather than tearing down and rebuilding the entire house. In fact, this is the way you would **have** to do it if you needed to continue to live in the house while changing it.

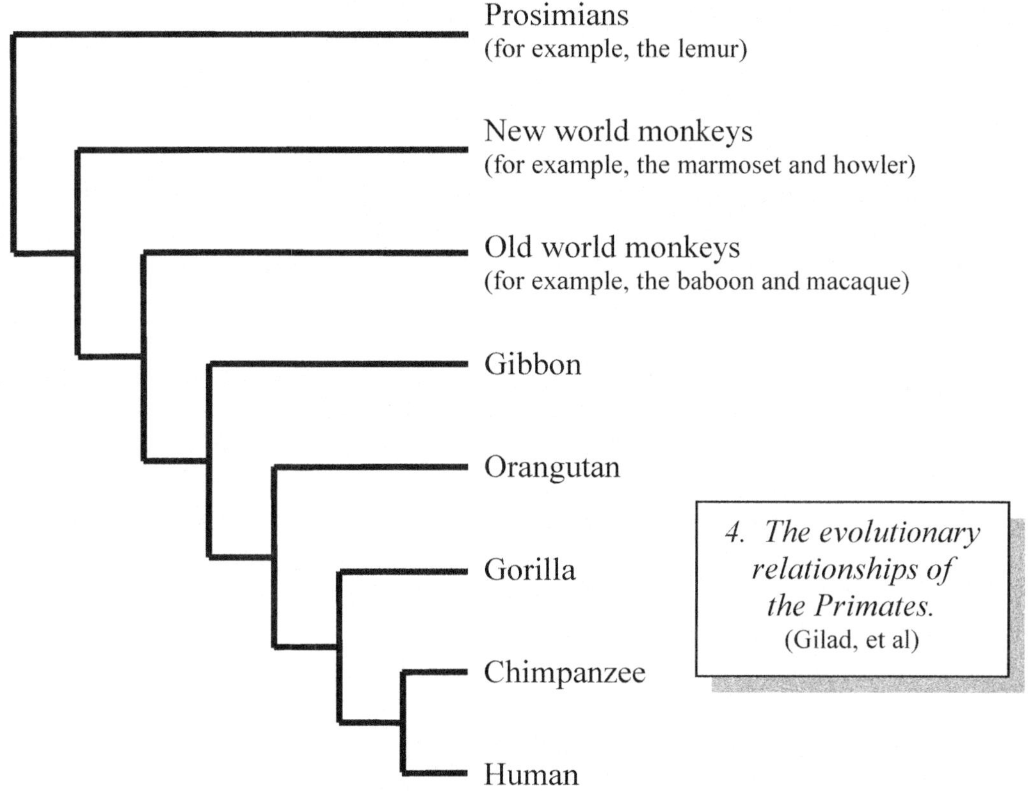

4. *The evolutionary relationships of the Primates.*
(Gilad, et al)

As Francois Jacob pointed out, "Natural selection…works like a tinkerer – a tinkerer who does not know exactly what he is going to produce but uses whatever he finds around him whether it be pieces of string, fragments of wood, or old cardboards; in short it works like a tinkerer who uses everything at his disposal to produce some kind of workable object…From an old bicycle wheel, he makes a roulette; from a broken chair the cabinet of a radio. Similarly evolution makes a wing from a leg or a part of an ear from a piece of jaw."[46] Something that initially evolved to serve a particular function, and then later was co-opted by natural selection for a different purpose is called an exaptation. Stephen Jay Gould wrote, "The best illustrations of adaptation by evolution are the ones that strike our intuition as peculiar or bizarre…Our textbooks like to illustrate evolution with examples of optimal design - nearly perfect mimicry of a dead leaf by a butterfly or of a poisonous species by a palatable relative. But ideal design is a lousy argument for evolution, for it mimics the postulated action of an omnipotent creator. Odd arrangements and funny solutions are the proof of evolution - paths that a sensible God would never treat but that a natural process, constrained by history, follows perforce."[47, 48, v] Darwin even wrote an entire book on orchids to argue that the structures used to insure fertilization by insects are jury-rigged by natural selection from available parts usually fitted for very different functions. "Although an organ may not have been originally formed for some special purpose, if it now serves for this end we are justified

---

 v   I hasten to point out that evolution does not rule out the existence of God, but it does rule out the Genesis account of creation.

in saying that it is specially contrived for it. On the same principle, if a man were to make a machine for some special purpose, but were to use old wheels, springs, and pulleys, only slightly altered, the whole machine, with all its parts, might be said to be specially contrived for that purpose. Thus throughout nature almost every part of each living being has probably served, in a slightly modified condition, for diverse purposes, and has acted in the living machinery of many ancient and distinct specific forms."[49]

Another source of evidence for evolution is the **vestigial structure**, an anatomical structure that is reduced and rudimentary compared to the same complex structure in other organisms. Darwin pointed out that "far from presenting a strange difficulty, as they assuredly do on the old doctrine of creation, [they] might even have been anticipated in accordance with"[50] the view of descent with modification. Vestigial features have been inherited from ancestors in which the structures were well developed and more important. Darwin wrote, "Organs now of trifling importance have probably in some cases been of high importance to an early progenitor, and after having been slowly perfected at a former period, have been transmitted to existing species in nearly the same state, although now of very slight use."[51] Though many vestigial organs have no function, complete non-functionality is not a requirement for vestigiality. Darwin wrote, "An organ may become rudimentary for its proper purpose, and be used for a distinct one… Rudimentary organs…are either quite useless, such as teeth which never cut through the gums, or almost useless, such as the wings of an ostrich, which serve merely as sails."[52]

Fossils or organisms that show the intermediate states between an ancestral form and that of its descendants, which are referred to as transitional or **intermediate forms**, support evolution. There are many examples of **transitional forms** in the fossil record, providing an abundance of evidence for change over time. A fossil, known as "Lucy"[vi], about 3 million years old and found at Hadar in Ethiopia, for example, exhibits a mixture of ape-like and human-like features.

Biogeography, the study of the past and present geographic distribution of plants and animals, supports evolution. The different species within any one region share similarities with others in the same region, rather than with other species living in similar environments in remote geographical regions. The geographic distribution of plants and animals cannot be explained by special creation, but makes sense from an evolutionary point of view. Each species originates in a certain place, becomes distributed by migrating, and evolves descendant species in the places to which it has migrated. In 1871, based on the fact that humans most resemble chimpanzees and gorillas, which live only in Africa, Darwin correctly predicted that fossils of our own earliest ancestors would be found there. "We are naturally led to enquire, where was the birthplace of man at that stage of descent when our progenitors diverged from the Catarhine stock? The fact that they belonged to this stock clearly shews that they inhabited the Old World; but not Australia nor any oceanic island, as we may infer from the laws of geographical distribution. In each great region of the world the living mammals are closely related to

---

vi   "Lucy" was the name given to the fossilized skeleton representing an individual of the species *Australopithecus afarensis.*

the extinct species of the same region. It is therefore probable that Africa was formerly inhabited by extinct apes closely allied to the gorilla and chimpanzee; and as these two species are now man's nearest allies, it is somewhat more probable that our early progenitors lived on the African continent than elsewhere."[53]

Comparative embryology supports evolution. Human embryonic development is similar to that of other vertebrates, more like that of other mammals than non-mammals, and most similar to that of other primates. In 1828, Ernst von Baer claimed that the more closely related any two species are, the more similar their development. Darwin argued that such homologies are an indication of descent from a common ancestor (see Topic 22).

Comparative molecular biology supports evolution. The pattern of biochemical similarities between humans and other mammals should and does agree with the pattern based on anatomical comparisons. On average, 98.77 percent of the DNA found in both humans and chimpanzees (*Pan troglodytes*) is identical and 1.23 percent is different.[54] Gorillas differ somewhat more, by about 2.3 percent, from humans or from either of the chimpanzees.[55] Thus, humans and chimpanzees are more closely related to one another than either is to the gorilla, or any other mammal, based on the molecular data.[56, 57, 58, vii] In fact, some scientists propose that "genus *Homo* should include three extant species and two subgenera, *Homo* (*Homo*) *sapiens* (humankind), *Homo* (*Pan*) *troglodytes* (common chimpanzee), and *Homo* (*Pan*) *paniscus* (bonobo chimpanzee)."[59]

The study of DNA has also revealed the existence of pseudogenes. A **pseudogene** is a damaged non-functional copy of a gene.[60] One kind of pseudogene arises when harmful mutations in a gene are not eliminated by natural selection because the function of the gene product has become unnecessary. For example, most mammals have a functional gene for producing the enzyme L-gulano-γ-lactone oxidase (GLO), which they need in order to synthesize ascorbic acid (vitamin C). However, guinea pigs and primates, including humans, are unable to synthesize ascorbic acid because they have a mutated GLO gene.[61, 62] The human and chimpanzee GLO pseudogenes are the most similar, followed by the human and orangutan genes, followed by the human and macaque (rhesus) monkey genes.[63] These facts are explicable by evolution: Assuming that the common ancestor of the primates had a diet rich in vitamin C, the inability to make it was of no consequence for survival and reproduction, so that there was no selective pressure against the defective gene. This kind of pseudogene can be considered a vestigial DNA sequence, since it is the relic of a gene whose function was important in ancestral species but has become unnecessary in the modern species. Such pseudogenes would not be expected if each species was independently created by an Intelligent Designer.[64]

A scientist is like a detective who must weigh the evidence, which sometimes completely excludes some suspects from consideration in the solution of a case, and in other instances strongly points to the identity of a particular culprit.[65] When fingerprints, blood, eyewitnesses, and surveillance cameras all agree, a detective can be confident that

---

[vii] This means that humans and chimpanzees probably share a common ancestor that lived about 5 to 6 million years ago, but **not** that humans evolved from chimpanzees!

he has solved the case. So it is with the evidence for evolution: The convergence of these disparate facts and independent lines of evidence strongly supports the reality of evolution.

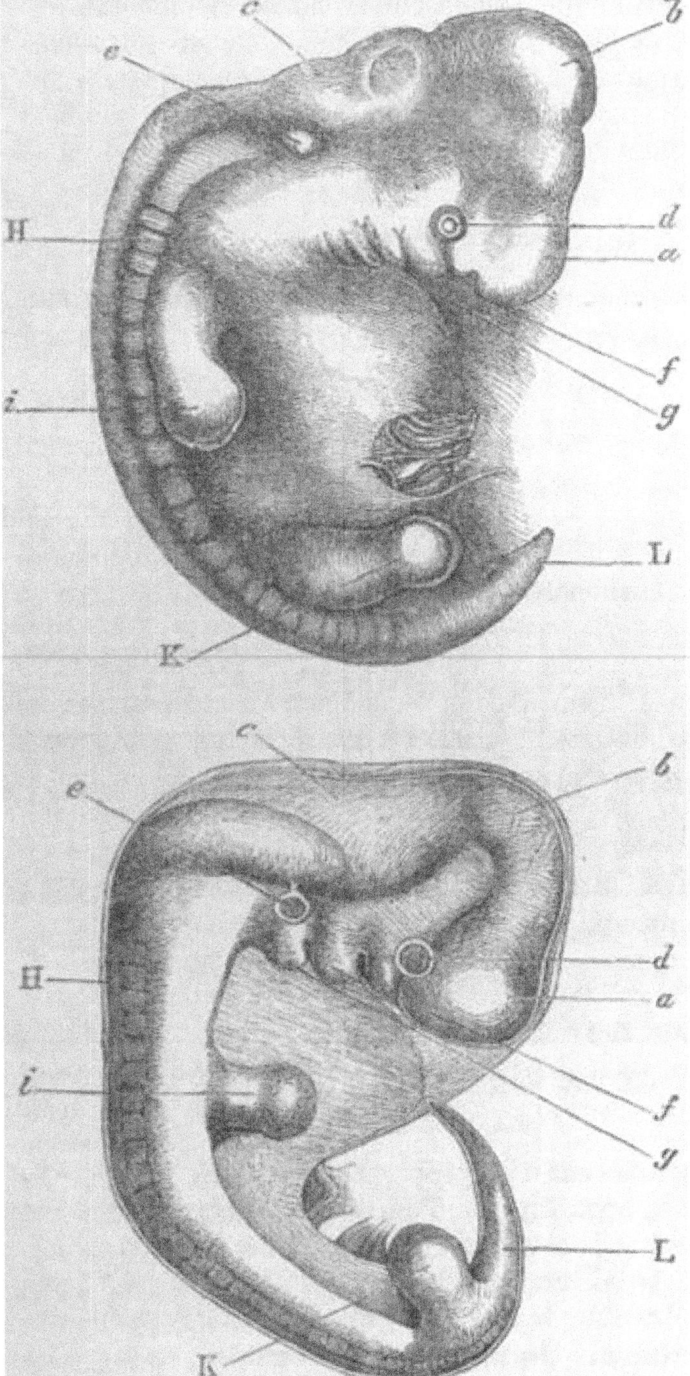

5. *Comparison of embryo of human (upper figure) and dog (lower figure).*
(Darwin, *The Descent of Man*, Figure 1)

a) Fore-brain, cerebral hemispheres, &c.
b) Mid-brain, corpora quadrigemina.
c) Hind-brain, cerebellum, medulla oblongata.
d) Eye.
e) Ear.
f) First visceral arch.
g) Second visceral arch.
h) Vertebral columns and muscles in process of development.
i) Anterior extremities.
k) Posterior extremities.
l) Tail or os coccyx.

TOPIC 5. GETTING YOUR BEARINGS

Human anatomy is the study of structure and its relation to function in the human body, while human physiology is study of the normal functioning of the human body. Structure and function are interrelated because a body part is able to perform a specific function due to its specific structure. Human anatomy can be studied in different ways:

- Surface anatomy is the study of anatomical landmarks on the surface of the body through visualization and palpation.
- Gross (macroscopic) anatomy is the study of structures that can be examined without using a microscope.
- Systemic (systematic) anatomy is the study of the structure of specific systems of the body, such as the digestive or cardiovascular systems. In systemic anatomy, the student considers one system at a time throughout the entire body. Thus, when the cardiovascular system is discussed, the student is presented with all the blood vessels from the head down to the big tow. The systemic approach is more suited to students whose biology backgrounds are not extensive and whose health career requirements are best fulfilled by a superficial exposure to anatomy. This is the approach used by this book.
- Regional anatomy is the study of specific regions of the body, such as the head or chest. In regional anatomy, the student examines all the systems of a particular region – its muscles and bone, blood supply, lymphatic drainage, and innervation. A good deal of attention is placed on the spatial relations of the various organs and structures in the region under consideration. The regional approach is better suited to students who have had advanced biology courses, such as vertebrate anatomy and embryology, and whose health career requirements are best fulfilled by an in-depth exposure to anatomy.
- Radiographic anatomy is the study of body structures that can be visualized with x-rays.
- Developmental anatomy is the study of structures that emerge from the time of the fertilized egg to the adult form.
- Embryology is the study of structures that emerge from the time of the fertilized egg through the eighth week in utero. (In studying other animals, embryology is part of developmental anatomy.)
- Histology is the study of the microscopic structures of tissues.
- Pathological anatomy is the study of structural changes (from gross to microscopic) associated with disease.

The human body is an organism that has several levels of structural organization:

- An organ system, which is a group of organs with a unique collective function.
- An organ, which is a structure formed by two or more kinds of tissues that has a distinctive function. Organs have definite anatomical boundaries and are visibly distinguishable from adjacent structures.
- A tissue, which is a group of similar cells and associated extracellular material that performs one or more specific functions.

- A ccll, which is the smallest unit of an organism capable of carrying out all the basic functions of life. Human cells are surrounded by a plasma membrane, contain a nucleus at some time in their lives, and are filled with a material called cytoplasm. Nothing less than a cell is considered to be alive.
- An organelle, which is a microscopic structure in the cytoplasm that is specialized to carry out individual functions of the cell.
- A molecule, which is a specific combination of individual atoms held together by covalent chemical bonds (see Topic 6).
- An atom, which is the smallest particle of matter with distinct chemical properties.

The major organ systems of the human body, their major components, and their broad functions are given in Table 3. The immune system is not really an organ system but a collection of disease-fighting cells that populate the lymphatic system and other organ systems. The principal functions of most organ systems overlap because some organs belong to two or more systems. For example, the male urethra is part of both the urinary system and the male reproductive system; the pharynx is part of both the respiratory system and the digestive system; and the pancreas is part of both the digestive system and the endocrine system. In the organism, all the systems function together to maintain homeostasis, which is defined as stability of the body's internal environment. However, internal conditions are not absolutely constant; rather they tend to fluctuate within a narrow range. For example, regardless of external conditions, the temperature of the body is maintained around 37° C (97° to 99° F).

The standard anatomical position is the position of the body universally used in anatomical descriptions: the body is erect, facing the observer with the head level and the eyes facing forward, the upper limbs are at the sides, the palms are facing forward, and the feet are on the floor and directed forward. Note that if a subject in anatomical position is facing the observer, the subject's left will be on the observer's right and vice versa. A body lying face down is in the prone position, and a body lying face up is in the supine position.

A plane is an imaginary flat surface that passes through the body or an organ. A section is a flat surface of the body or an organ revealed by cutting along a specific plane. A sagittal plane is a vertical plane that divides the body or organ into left and right portions. There are two types: midsagittal and parasagittal. A midsagittal (or median) plane passes through the length of the sagittal suture of the skull and thereby divides the body into equal right and left halves. A parasagittal plane is any plane parallel to and displaced from the midsagittal, which therefore divides the body into unequal right and left portions. A frontal (or coronal) plane is a vertical plane that divides the body or organ into anterior (front) and posterior (back) portions. A transverse plane is a horizontal plane that divides the body or organ into superior (upper) and inferior (lower) portions. Any sagittal, frontal, or transverse plane is perpendicular to the other two. An oblique plane is a plane that is not in any of the sagittal, frontal or transverse planes.

Table 3. The Human Organ Systems

Organ Systems		Organs	Functions
	Respiratory	Nose, trachea and lungs.	Input and output.
	Digestive	Mouth, esophagus, stomach, small and large intestines, liver, and pancreas.	
	Integumentary	Skin, hair, nails, sebaceous glands, and sudoriferous glands.	Protection, support, and movement.
Musculoskeletal	**Skeletal**	Bones, joints, and their associated cartilages.	
	Muscular	Muscles composed of skeletal muscle tissue.	
Circulatory	**Cardiovascular**	Heart, blood vessels, and blood.	Fluid transport.
	Lymphatic/ Immune	Lymph, lymph nodes, lymphatic vessels, spleen, and leukocytes, including T-lymphocytes and B-lymphocytes.	Fluid transport. Defense.
Neuroendocrine	**Nervous**	Brain, spinal cord, and peripheral nerves.	Internal communication and integration.
	Endocrine	Glands that secrete hormones.	
Urogenital	**Urinary**	Kidneys, ureters, urinary bladder, and urethra.	Input and output.
	Reproductive, Male	Testes, ductus (vas) deferens, seminal vesicles, prostate gland, and penis.	Reproduction.
	Reproductive, Female	Ovaries, oviducts, uterus, vagina, and mammary glands.	

Directional terms are used to locate one body structure relative to another. Because humans walk bipedally in an upright stance, some directional terms have different meanings for humans than they do for animals that walk on four legs. For example, anterior refers to the region of the body that leads the way in locomotion and posterior refers to the region that comes last. For a quadrupedal (four-legged) animal, the head end is anterior and the tail end is posterior, while the belly side of the body is referred to as ventral and the side nearest the spine as dorsal. However, for a human, the region of the chest and abdomen is anterior and the region of the back and buttock is posterior. Thus, anterior has the same meaning as ventral and posterior has the same meaning as dorsal for a human, but not for a cat. Other directional terms include:

- Superior means above or upper.
- Inferior means below or lower.
- Medial means closer to the midsagittal plane.
- Lateral means farther from the midsagittal plane.
- Proximal means closer to a point of origin or attachment.
- Distal means farther from a point of origin or attachment.
- Superficial means closer to the surface of the body or organ.
- Deep means farther from the surface of the body or organ.

- Ipsilateral means on the same side.
- Contralateral means on the opposite side.
- Central means near or toward the midline of the body.
- Peripheral means away from the midline or center of the body.

The human body can be divided into a number of regions. The axial region of the body consists of the: head (cephalic region), consisting of the skull (cranial region) and the face (facial region); neck (cervical region); and trunk, including the thoracic region (chest) above the thoracic diaphragm and the abdominal region below the thoracic diaphragm. Other regions that are commonly referred to in anatomical descriptions are as follows:

- Axillary region (armpit), which is the depression between the arm and the trunk.
- Antecubital region, which is the depression at the front of the elbow.
- Popliteal region, which is the depression at the back of the knee.
- Gluteal region, which is the region of the buttocks.
- Pudendal (genital) region, which is the region around and including the external sex organs.
- Anal region, which is the area round and including the anus.
- Perineal region, which is the diamond-shaped region between the thighs that includes the anal and pudendal regions.

The abdominal region can be further divided into nine regions (used more often by anatomists) or four quadrants (used more often by clinicians). The nine regions are defined by four lines, two horizontal and two vertical, that intersect like a tic-tac-toe grid. The three lateral regions on each side of this grid, from upper to lower, are the hypochondriac, lumbar (lateral abdominal), and iliac (inguinal). The three medial regions, from upper to lower, are the epigastric, umbilical, and hypogastric (pubic). The four quadrants are defined by two perpendicular lines, one horizontal and the other vertical, that intersect at the umbilicus (navel): right upper quadrant, left upper quadrant, right lower quadrant, and left lower quadrant.

The appendicular region of the body consists of the upper extremities and lower extremities (or limbs). Each upper extremity includes the: brachial region or brachium (arm); antebrachial region or antebrachium (forearm); carpal region or carpus (wrist); manual region or manus (hand); and digits (fingers). Each lower extremity includes the: femoral region (thigh); crural region or crus (leg); tarsal region or tarsus (ankle); pedal region or pes (foot); and digits (toes).

A body cavity is a closed space within the body that helps protect, separate, and support internal organs, called the viscera. Note that the alimentary canal and respiratory, urinary, and reproductive tracts are not body cavities because they are open to the outside and do not contain internal organs. The two major cavities of the body are the dorsal and the ventral.

The dorsal body cavity is near the posterior surface of the body and consists of the cranial cavity, which is enclosed by the cranium and contains the brain, and the vertebral cavity or canal, which is enclosed by the vertebral or spinal column (backbone) and contains the spinal cord. The ventral body cavity is near the anterior side of the body and arises from the intraembryonic coelom of the embryo. It consists of the thoracic cavity, which is superior to the thoracic diaphragm and contains the lungs, heart, and the thymus, and the abdominopelvic cavity, which is inferior to the thoracic diaphragm. The abdominopelvic cavity in turn can be divided into the abdominal cavity, which contains the spleen, most of the digestive organs, the kidneys, and ureters, and the pelvic cavity, which contains the distal part of the large intestine, the urinary bladder and urethra, and the reproductive organs.

TOPIC 6. THE CHEMISTRY OF LIFE

Matter is anything that has mass and occupies space. There are two types of matter: pure substances and mixtures. A pure substance has a constant composition wherever it occurs and is either an element or a chemical compound. A mixture contains two or more (pure) substances whose proportions vary among mixtures and sometimes among regions within the same mixture.

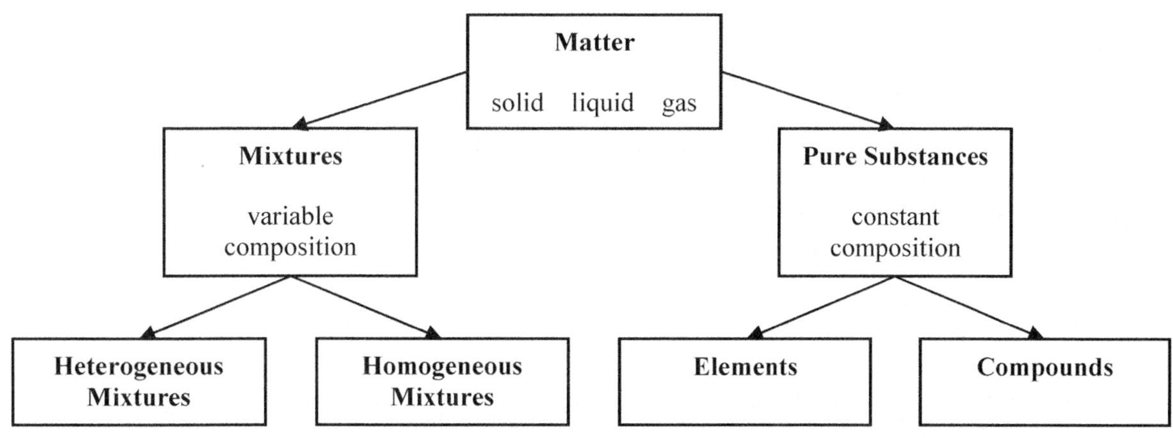

6. The classification of matter. Physical methods can be used to separate mixtures into pure substances or to blend pure substances into mixtures.

According to atomic theory, matter is fundamentally composed of atoms. Atoms, in turn, are composed of protons, neutrons, and electrons. A proton (p^+) has a single positive charge and a neutron (n^0) has no charge; they each weigh about 1 atomic mass unit (amu). An electron (e^-) is a tiny particle with a single negative charge and a very low mass, about 0.0005 amu. The number of protons and electrons in an atom are equal because the atom, by definition, has no net charge. Protons and neutrons are located in the nucleus at the center of the atom, and electrons are constantly moving in nearby regions called electron clouds.

An **element** consists of all atoms with the same number of protons, called its atomic number. There are 92 naturally occurring elements on Earth, and over 95 percent (by weight) of the human body consists of six of them: carbon, nitrogen, oxygen, hydrogen, calcium, and phosphorus. Atoms of a given element usually vary in their numbers of neutrons. An isotope consists of all atoms of an element with the same number of neutrons. In other words, an isotope consists of all atoms with the same number of neutrons and protons, called its atomic mass. Most elements have several isotopes. Some isotopes are unstable; their atoms over time decay (break down) into smaller, more stable particles, releasing radiant energy in the process. Thus, they are called radioisotopes.

A **chemical compound** consists of two or more elements combined in the same fixed proportions. The atoms in the compound are held together by chemical bonds arranged in precise order, normally unchanging over time, and with a relatively fixed geometric relationship. The fixed proportions of elements in a compound are expressed by the subscripts in its chemical formula. For example, water is a compound consisting of hydrogen and oxygen. For each oxygen atom, there are two hydrogen atoms, so the chemical formula of water is H_2O.

Atoms combine with each other to form compounds in two major ways. One of these ways is by the transfer of one or more electrons from one atom to another. For this to occur, the two atoms must belong to different elements. Electron transfer produces electrically charged particles called ions. The process of an individual atom gaining or losing an electron to form an ion is called **ionization**. The atom that has lost an electron becomes a positively-charged ion, called a cation, and the one that has gained an electron becomes a negatively-charged ion, called an anion. Compounds composed of ions are called **ionic compounds**. An ion is represented by the chemical symbol for the element followed by a plus sign or a minus sign for each electron lost or gained, respectively. For example, when sodium and chlorine react, each sodium atom loses an electron and each chlorine atom gains an electron, producing sodium chloride (NaCl or table salt). Thus, sodium chloride is an ionic compound composed of sodium ions (Na^+) and chlorine ions (Cl^-). Strong electrical attractions, called ionic bonds, between the cations and anions of an ionic compound hold them together in such a way that their charges tend to balance one another, forming an electrically-neutral **crystal**.

If a crystal is placed in water, the water molecules, which are somewhat electrically charged themselves (see below), attract each ion away from the other ions in the crystal. The sum of the weak charges on the water molecules exceeds the strength of the ionic bond. The separation of ions from one another that occurs when a crystal dissolves in water is called **dissociation**. Ions are extremely important in many of the processes of life. Numerous ions must be present in just the right amounts inside cells, in tissue fluids, and in the blood for the human body to function properly. For example, calcium ions are essential for triggering the contraction of muscle cells. Without calcium, the heart will fail to function.

The other major way compounds are formed is by the sharing of one or more pairs of electrons between atoms. This sharing of electrons holds the atoms together to form discrete electrically-neutral particles called **molecules**, and each pair of electrons shared between two atoms is called a covalent bond. Compounds composed of molecules are called **molecular compounds**. Water (H_2O) is an example of a molecular compound; each molecule consists of one oxygen atom and two hydrogen atoms. Unlike ionic bonds, a covalent bond can form between two atoms of the same element. In fact, some molecules contain atoms of only one element and are referred to as the molecular form of that element. An example is molecular oxygen (O_2), in which each molecule consists of two oxygen atoms.

In some cases, two atoms that are covalently bonded share their electrons equally. For example, a molecule of methane (CH_4) consists of a carbon atom that shares electrons with four hydrogen atoms. That sharing is nearly equal; in other words, the electrons in each covalent bond spend the same amount of time around the carbon atom and around the hydrogen atom. Therefore, each covalent bond is said to be nonpolar; that is, it lacks distinctive poles with opposite electrical charges. In addition, since no part of the surface of the molecule has an electrical charge that is any different from any other part, such a molecule is said to be **nonpolar**.

When oxygen shares electrons with hydrogen, the result is somewhat different. Since the oxygen atom has two more protons in its nucleus than carbon does, it has a stronger attraction for electrons than carbon. As a result, the electrons that the oxygen is sharing with the hydrogens are not shared equally. They spend more time around the oxygen atom than around the hydrogen atom – so much so that they overbalance the positive charges of the oxygen nucleus. This gives the oxygen atom a distinct, although weak, negative charge, indicated by the Greek lowercase letter delta and a minus sign (δ^-). The electrons don't spend enough time around the hydrogen atom to quite neutralize the positive charge of the hydrogen's nucleus, so the hydrogen atom acquires a distinct, although weak, positive charge designated as delta positive (δ^+). These delta charges are only about one-fifth as strong as the full unit of charge on each sodium and chlorine ion. Such a covalent bond is said to be **polar** because it has two distinctive regions, or poles, with opposite electrical charges. For example, a molecule of water consists of an oxygen atom that shares electrons with two hydrogen atoms, but the sharing is unequal. Not only is each covalent bond polar, but the molecule as a whole is polar because it is asymmetrical.   The hydrogen atoms are not on opposite sides of the oxygen atom but are more toward one side than the other. The result is that one side of the water molecule is delta positive and the opposite side is delta negative. This polarity of water and similar molecules has a profound effect on their behavior.

Water molecules are rather strongly held together by the attraction between the delta positive charges on the hydrogen atoms of one water molecule and the delta negative charges on the oxygen atoms of other water molecules. As a result of this attractive force, or bond, water remains a liquid and does not rapidly become a gas, as methane does, unless additional heat energy is put into it. Such bonds between hydrogen atoms in a polar molecule and a negatively charged atom in some other molecule are called

hydrogen bonds.

When a solid is dissolved in a liquid, the result is a type of mixture called a **solution**, in which the solid is called a **solute** and the liquid is called the **solvent**. Polarity and hydrogen bonding make water not only a good solvent, but the best. It dissolves more different substances than any other solvent known. This is because so many other substances consist of ions or polar molecules, the electrical charges of which attract water molecules and cause these substances to stay in solution. Even nonpolar molecules dissolve to some extent if they are small, such as molecular oxygen (O_2) and carbon dioxide (CO_2). Substances that dissolve in water are said to be **water-soluble**; substances that do not are **water-insoluble**. The ability of water to dissolve so many substances is important because when dissolved, they move about and collide, thereby facilitating chemical reactions.

One interesting feature of water and many other covalent compounds is that they, too, can dissociate into ions. Unlike ionic compounds, such as sodium chloride, they are not ionized before they dissociate; they accomplish ionization and dissociation simultaneously. When water dissociates, one of the hydrogen nuclei loses its electron to the oxygen atom and becomes a **hydrogen ion** (really just a proton except in the rare isotopes of hydrogen), while the oxygen and the other hydrogen atom become a **hydroxide ion**. The hydrogen ion has a full unit of positive charge and is represented as H^+. Since the hydroxide ion has gained an electron from the departed hydrogen, it has a full unit of negative charge and is represented as OH^-. On average, as one hydrogen ion combines with a hydroxide ion to form a water molecule, another water molecule dissociates to replace the hydrogen ion and hydroxide ion in solution. As a result, very few of the water molecules are dissociated at any one time – in fact, only about one in 550 million.

Certain compounds dissociate in such a way that they release hydrogen ions without releasing hydroxide ions. Such substances are called **acids**. For example, hydrochloric acid (HCl) is an ionic compound that dissociates easily in water to produce a solution of hydrogen ions (H^+) and chloride ions (Cl^-). On the other hand, a **base** or **alkali** removes hydrogen ions from solution. For example, sodium hydroxide (NaOH) is a molecule that dissociates easily in water, forming a solution of sodium ions and hydroxide ions. These hydroxide ions then combine with the hydrogen ions in solution to form water molecules. A compound that dissociates to produce ions other than hydrogen ions or hydroxide ions is called a **salt**, such as sodium chloride (NaCl).

The acid or basic strength of a solution may be expressed in terms of a number called the **pH** of that solution. The pH scale ranges from 0 at the acid end to 14 at the basic end. Each change of one pH unit indicates a tenfold increase or decrease in hydrogen ion concentration and the opposite tenfold change in hydroxide ion concentration. Pure water has a pH of 7, which means it has equal (although extremely small) concentrations of hydrogen ion and hydroxide ion. Most fluids in the human body have a pH close to 7, although stomach acid usually has a pH of 1.

The sum of all the chemical reactions that occur within the body is called its metabolism. Metabolism is customarily divided into two parts: catabolism and anabolism. Anabolism consists of reactions in which larger organic molecules are synthesized from smaller ones with the consumption of energy. Catabolism consists of reactions in which complex organic compounds are broken down into simpler ones with the release of energy. Chemical reactions in which molecular oxygen is a reactant are said to be **aerobic** and those in which it is not are **anaerobic**.

Most molecules found in living things contain hundreds or thousands of atoms, so they are called biological macromolecules. Cells construct macromolecules by covalently joining smaller molecules, called subunits. Many macromolecules are polymers, which consist of many similar subunits joined together to form a long chain. The major types of macromolecules found in all living things, including the human body, are carbohydrates, lipids, proteins, and nucleic acids.

"Hydrate" refers to something combined with water, and carbohydrates have approximately two hydrogen atoms and one oxygen atom (amounting to one water molecule) for every carbon atom. Thus, the general formula for carbohydrates is $(CH_2O)_n$, where n can be almost any number. For example, the n of glucose is 6; multiplying everything within the parentheses by 6, the chemical formula of glucose is $C_6H_{12}O_6$. All carbohydrates attract water molecules and are therefore said to be **hydrophilic** ("water-loving").

There are three types of carbohydrates: monosaccharides, disaccharides, and polysaccharides. A monosaccharide is a single subunit, such as glucose. Other monosaccharides include fructose and galactose. A disaccharide consists of two subunits joined together, such as sucrose or lactose. Sucrose is formed by combining a glucose molecule and a fructose molecule, and lactose is formed by combining a glucose molecule and a galactose molecule. Both monosaccharides and disaccharides are also known as simple sugars. Simple sugars are water-soluble and sweet to the taste. A polysaccharide, also known as a complex carbohydrate, consists of many monosaccharides joined together in a long chain. For example, glycogen is a polymer of glucose. Polysaccharides are water-insoluble and are not sweet to the taste.

All lipids are greasy and dissolve poorly or not at all in water. This is because, for the most part, they do not attract water molecules and instead prefer their own company. Molecules like this are said to be **hydrophobic** ("water-fearing"). There are three types of lipids, none of which are polymers: triglycerides, phospholipids, and steroids. Triglycerides, commonly called fats and oils, are used for long-term energy storage and do not dissolve in water. Phospholipids are a major component of cell membranes. One end of the molecule, called the "head," is hydrophilic and water-soluble but the other end, called the "tail," is hydrophobic and water-insoluble. Molecules like this are said to be amphipathic. All steroids, including cholesterol and many hormones, have a basic structure of interconnected carbon rings and do not dissolve in water.

The fundamental subunits of proteins are amino acids. Cells use twenty different kinds of amino acids. Two amino acids covalently joined together are called a dipeptide, while a long chain (polymer) of amino acids is called a polypeptide. A protein is composed of one or more polypeptides. Proteins are used as catalysts (enzymes) for chemical reactions, as hormones, or as parts of structures.

7. The basic ring structure of all steroids.

The subunits of nucleic acids are nucleotides. A polymer of nucleotides is called a polynucleotide. Nucleic acids occur in two forms in the cell. Deoxyribonucleic acid (DNA), which is found in the chromosome, stores genetic information. Ribonucleic acid (RNA) is involved in making proteins using the genetic information in DNA. The nucleotides of DNA and RNA are slightly different.

Finally, another molecule that deserves special mention is adenosine triphosphate (ATP). ATP is a modified RNA nucleotide that is the immediate source of energy in the cell for all biological processes. Cells use energy they get from the breakdown of glucose to combine adenosine diphosphate (ADP) with inorganic phosphate (P_i) to make ATP. When ATP is broken down to ADP and inorganic phosphate, energy is released that the cell can then use to do work.

TOPIC 7. CELLS

The human body consists of 100 trillion cells and about 200 different cell types. Each cell is either a somatic (body) cell or a germ-line cell. Somatic cells are diploid, containing 46 chromosomes, and are by far the most common. Mutations – permanent changes in DNA – in a somatic cell may lead to cancer in the body, but the mutations cannot be passed on to the next generation. Germ-line cells are also diploid but are involved in sexual reproduction. They produce the gametes, the eggs (female) or sperm (male), which are haploid, containing only 23 chromosomes. Thus, mutations in a germ-line cell can be inherited by the offspring.

The shapes of cells vary widely:

- Squamous cells, such as those lining the esophagus and covering the skin, are thin and flat.
- Cuboidal cells, such as liver cells, are about as tall as they are wide.
- Columnar cells, such as cells lining the intestines, are significantly taller than wide.
- Discoid cells, such as erythrocytes, are disc-shaped.
- Spheroid to ovoid cells, such as egg cells and adipocytes, are round to egg-shaped, respectively.
- Fusiform cells, such as the contractile cells of smooth muscle tissue, are thick in the middle and tapered toward the ends – often described as "spindle-shaped."
- Stellate cells, such as nerve cells with multiple extensions, are star-shaped.

All living cells on Earth must be bathed in a watery solution because the first cells evolved in an aquatic environment. Fluid outside the cells of the body is called extracellular fluid and is found in several locations. The extracellular fluid that occupies the spaces between the cells of tissues is called tissue fluid, interstitial fluid, or intercellular fluid; in lymphatic vessels, it is called lymph. The extracellular fluid in blood vessels is called plasma.

A typical human cell consists of two major parts: the plasma membrane and the cytoplasm. The cytoplasm consists of the cytosol in which are embedded three kinds of structures: organelles, the cytoskeleton, and inclusions. Cytosol, also known as intracellular fluid, is an aqueous solution – water is the solvent. Organelles are intracellular structures, usually enclosed by a membrane, with characteristic shapes and functions. The cytoskeleton is an intracellular system of microtubules (thin cylinders composed of the protein tubulin) and protein filaments that physically supports the cell, helps organize its contents, moves substances through the cytoplasm, and contributes to movements of the entire cell. Inclusions are aggregates of organic molecules in the cytoplasm, such as granules of melanin in skin cells, glycogen in liver cells, and triglycerides in fat cells.

The plasma membrane serves as the boundary between the inside of the cell and the outside. About 40 to 50 percent of the membrane consists of lipids and 50 to 60 percent consists of proteins. Membrane lipids are mostly phospholipids, with some cholesterol and glycolipids. The phospholipids are arranged in two layers, referred to as the lipid bilayer, with their hydrophobic tails directed toward the center of the membrane and their hydrophilic heads directed outward.

Membrane proteins can be divided into two structural groups: integral proteins and peripheral proteins. Integral proteins are proteins that are directly incorporated within the lipid bilayer. They may completely span the lipid bilayer (transmembrane proteins) or be embedded mostly on one side of the membrane. Integral proteins can be extracted from cell membranes only by methods that use detergents, which disrupt the lipid bilayer. Peripheral proteins, which are usually bound to exposed regions of integral proteins, are more loosely associated with membrane surfaces. They can be easily removed from the membrane with salt solutions without disrupting the structure of the lipid bilayer.

There are several functional types of membrane proteins. Carrier proteins are transmembrane proteins that permit movement of ions and molecules into or out of the cell. They can be divided into two types. Channel proteins function as "gates", which can open and close and allow a specific substance to move in either direction. Transporter proteins function as "pumps" that move a substance in only one direction across the membrane. Receptor proteins bind to chemical messengers on the outside of the cell, such as hormones and neurotransmitters, or are involved in cell-cell adhesion or other activities. Cell-identity markers include the ABO blood type markers and the major histocompatability (MHC) antigens.

At body temperature in humans, the lipid bilayer is a liquid with the consistency of olive oil. Some membrane proteins are anchored in place by the cytoskeleton. Other proteins and all phospholipids can move laterally in the bilayer, although they cannot change their orientation with regards to the outside and inside of the cell. This concept of proteins changing position in a liquid phospholipid bilayer is called the "fluid-mosaic model" of membrane structure.

All living cells have an electrical difference (measured in volts), called the resting membrane potential, across their plasma membrane. In other words, the membrane is polarized – its inner, cytoplasmic side is negatively charged with respect to its outer, extracellular side. Sufficient stimulation of a nerve cell or a muscle cell can cause a rapid but transient polarity reversal at a point in the plasma membrane, so that the plasma membrane is positively charged on its inside surface and negatively charged outside. Thus, muscle cells and nerve cells (neurons) are said to have the property of electrical excitability. The sequence of electrical changes resulting in this localized polarity reversal in the plasma membrane of the cell is called an action membrane potential or simply an action potential. In muscle cells, these action potentials result in contraction; in nerve cells, they result in the rapid transmission of signals, called nerve impulses, to other cells.

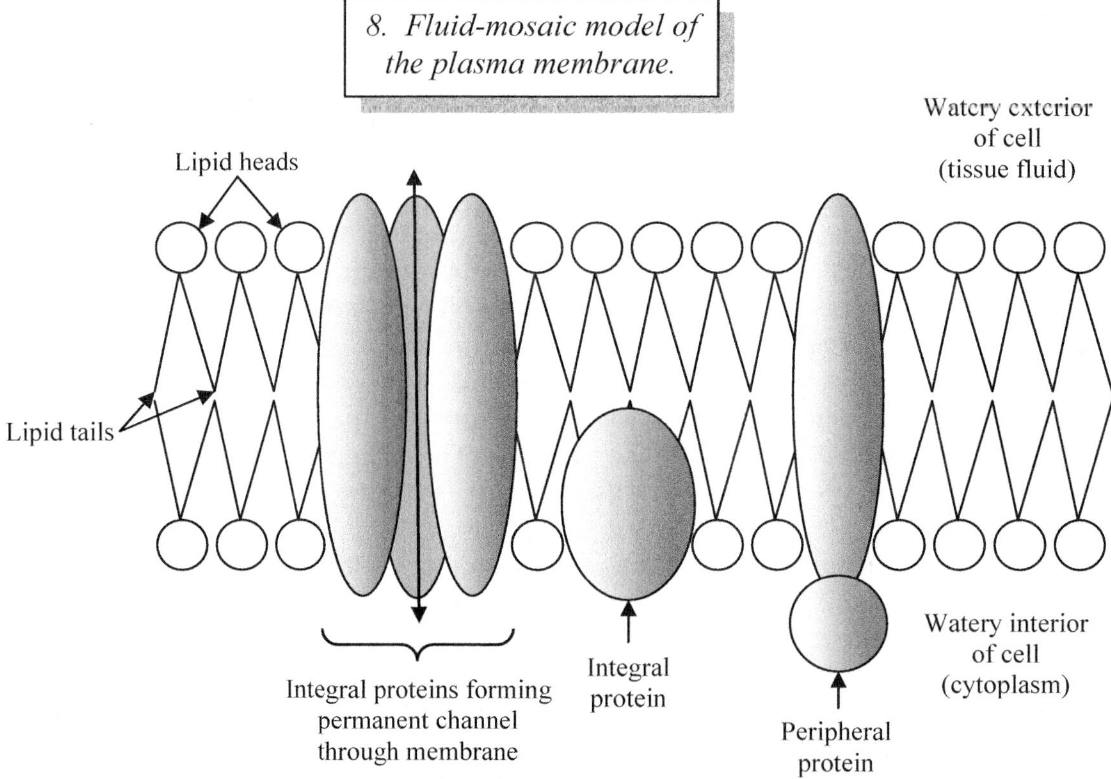

The chief function of the plasma membrane is to regulate membrane transport, which is the movement of substances across the membrane into or out of the cell. This ability to permit passage of certain substances and to reject others on the basis of size, lipid solubility, electrical charge, or affinity for carrier proteins in the membrane is called selective permeability or semi-permeability.

Substances can pass through the plasma membrane by passive or active processes. In passive processes, the energy or force for membrane transport is supplied by the extracellular environment. These processes involve either diffusion or filtration. Diffusion is the net movement of a solute through a solution from where it is more concentrated to where it is less concentrated. The difference in concentration between two regions of a solution is called a concentration gradient. As it progresses, diffusion will reduce and may eventually eliminate the concentration gradient. The diffusion of a solute across a semipermeable membrane is called dialysis, while the diffusion of water (the solvent) but not solute across a semipermeable membrane is called osmosis. Thus, in either dialysis or osmosis, there must be a concentration gradient across the membrane. Channel proteins are often required for dialysis to occur. Filtration is the net movement of solute and solvent through a semipermeable membrane from a region of higher fluid pressure to a region of lower fluid pressure. The difference in pressure across the membrane is called a pressure gradient.

In active processes, the energy for membrane transport is supplied by the cell. Active processes include active transport and vesicular transport. Active transport requires a transporter protein in the plasma membrane to move a small solute (usually an ion) into or out of the cell. Since the substance is moved from where it is less concentrated to where it is more concentrated, active transport **creates** a concentration gradient. Therefore, the cell must do work. Vesicular transport is an energy-requiring process in which a bulky substance, such as a bacterium or a large number of molecules, is moved into or out of the cell by the formation or fusion of vesicles. A vesicle is a small, membrane-enclosed sac within a cell that forms by budding off from an existing membrane. In exocytosis, materials move out of a cell when the vesicle that contains them fuses with the plasma membrane. In endocytosis, materials move into a cell within a vesicle formed from the plasma membrane. In either case, active and/or passive transport is still required to get material into or out of the vesicles within the cell.

Phagocytosis is a form of endocytosis in which the material being moved into the cell is a relatively large solid, such as a virus or whole bacterium. A cell that performs phagocytosis is called a phagocyte or a phagocytic cell.

Many cells have surface extensions called microvilli, cilia, and flagella. Microvilli (singular: microvillus) are extensions of the plasma membrane that serve primarily to increase a cell's surface area. Since they are individually only about 1 to 2 μm in length, they appear under the light microscope as a fringe called the brush border on the cell surface. Examples of cells with microvilli are the epithelial cells of the intestines and renal tubules. Microvilli give these cells, which are specialized for absorption, about 15 to 40 times as much absorptive surface area as they would have without microvilli.

Flagella (singular: flagellum) and cilia (singular: cilium) are hair-like projections of the plasma membrane that each enclose a bundle of parallel microtubules. Flagella are long and few. The sperm cell is the only human cell with a flagellum, which acts like a propeller. Cilia are short, about 7 to 10 μm in length, and numerous. Many human cells have them, such as those that line the respiratory tract. Cilia move materials past the surfaces of these cells like a conveyer belt.

Organelles can be classified into two types: those surrounded by one or two layers of membrane, therefore referred to as membranous organelles, and those not surrounded by membrane, called nonmembranous organelles. Membranous organelles include the nucleus, mitochondria, endoplasmic reticulum, Golgi complex, lysosomes, and peroxisomes. Nonmembranous organelles include the ribosomes, centrosome, and centrioles.

- The nucleus contains the cell's genetic instructions in the form of molecules of DNA. Each DNA molecule is coated with protein, forming a compact chromosome during cell division or elongated chromatin the rest of the time. A single molecule of DNA contains thousands of genes, which determine the heritable traits of the individual. The nucleus is enclosed by two membranes called the nuclear envelope, which are fused in many places to form openings called nuclear pores. At maturity, some cells have no nucleus (for example, erythrocytes) or multiple nuclei (for example, skeletal muscle fibers).
- Mitochondria (singular: mitochondrion) are double-membraned organelles where large quantities of ATP are produced by breaking down food molecules, such as glucose, to carbon dioxide and water. These aerobic chemical reactions are known as cellular respiration.
- The endoplasmic reticulum (ER) is a network of interconnected membrane-enclosed passageways in the vicinity of the nucleus. There are two basic types of ER. Rough ER is studded with ribosomes that manufacture proteins to be delivered to lysosomes or to the plasma membrane or to be secreted. Smooth ER lacks ribosomes and is the site of lipid synthesis, detoxification, storage of calcium in muscle cells, and other processes.
- The Golgi complex (or Golgi apparatus or Golgi body) is a series of flattened, membrane-enclosed sacs stacked upon each other (pancake-style). Here, proteins from the rough ER and lipids from the smooth ER are received. These proteins and lipids are modified, sorted, and packaged into vesicles for delivery to lysosomes or to the plasma membrane or for secretion. Hence, the Golgi complex is sometimes nicknamed the "traffic director" or "post office" of the cell.
- Lysosomes are vesicles that contain numerous digestive enzymes to be secreted (for example, into the stomach) or to be used for intracellular digestion.
- Peroxisomes are vesicles that are the site of metabolic processes that produce and subsequently break down hydrogen peroxide. They detoxify poisons in liver cells and kidney cells.
- Ribosomes are structures where proteins are made in the cell. These sites of protein synthesis are sometimes nicknamed the "protein factories" of the cell.
- The centrosome, located near the nucleus, contains a pair of centrioles. Centrioles are short cylindrical organelles composed of microtubules. They are important in cell division and the development of cilia and flagella.

The cell cycle consists of the sequence of events leading from the end of one cell division to the end of the next cell division. The stage between cell divisions is called interphase, during which growth and DNA replication occur. The cell spends most of its life in interphase.

Cell division can be somatic or reproductive. Somatic cell division produces two genetically identical daughter cells from one parent cell. It consists of mitosis and cytokinesis. Mitosis is a division of the nucleus that produces two identical daughter nuclei, each containing 46 chromosomes. Cytokinesis is a cytoplasmic division resulting in the formation of two daughter cells, each with its own nucleus. Reproductive cell division produces up to four genetically different daughter cells called gametes from one germ-line cell by a process called meiosis. Meiosis consists of two successive cell divisions without any intervening interphase. The first division is called meiosis I or the reduction division. The second division is called meiosis II or the equatorial division. The male gametes, or sperm, and the female gametes, or eggs, each have 23 chromosomes instead of the usual 46.

Cell junctions are points of contact between plasma membranes of adjacent cells. Three structural types are especially significant:

- A tight junction is the fusion of the plasma membranes of adjacent cells, acting mostly as a barrier to the movement of substances between the cells.
- A desmosome is an anchoring junction, acting like a spot weld.
- A gap (communicating) junction is a protein assembly between cells, forming channels that connect the cytoplasm of the adjacent cells.

TOPIC 8. AN OVERVIEW OF TISSUES

All tissues consist of two major components: cells and the nonliving extracellular material or matrix. The matrix includes the extracellular fluid and often contains protein fibers. The four basic types of tissue in the human body are:

- Epithelial tissue, which has only a small amount of extracellular matrix. Its principal functions are lining body surfaces exposed to the outside and also glandular secretion.
- Connective tissue, which has abundant extracellular matrix. It has several and diverse principal functions.
- Muscle tissue, which has a moderate amount of extracellular matrix. Its principal function is movement.
- Nervous tissue, which has no extracellular matrix. Its principal function is the transmission of nerve impulses.

All tissues develop from three primary germ layers: ectoderm, endoderm, and mesoderm. Epithelial tissues are derived from all three and nervous tissues develop from ectoderm.

Mesoderm contributes to an embryonic connective tissue called mesenchyme[viii], from which develops all permanent connective tissues, as well as most muscle.

TOPIC 8.1. EPITHELIAL TISSUE

Epithelial tissues are composed of closely packed polyhedral (when viewed from above) cells with very little extracellular matrix. Adhesion between these cells is strong, forming continuous sheets in either single or multiple layers. Epithelial tissues are avascular, but they almost always lie on a layer of connective tissue containing blood vessels with which they exchange substances via diffusion. Most epithelial tissues receive a rich supply of sensory nerve endings. Epithelial tissue has a high capacity for cell division in order to replace cells lost due to abrasion and injury.

Epithelial cells are attached to underlying connective tissue by a thin extracellular layer called the basement membrane[ix]. The basal surface of the cell is the surface that is attached to the basement membrane, and the lateral surfaces are the sides of the cell. The surface of the cell that faces the exterior or a lumen is called the apical surface. A lumen (plural: lumina) is a space within a tubular structure such as a blood vessel or intestine. Microvilli arise from the apical surfaces of most cells, thereby greatly increasing their surface area. They may be short or long (but never as long as cilia) and range in number from a few to many.

Epithelia can be structurally classified based on cell shapes and the number of layers present. Simple epithelium contains only one layer of cells, while stratified epithelium contains more than one layer of cells and is classified according to the shape of cells in its **surface** layer. This gives rise to the following types of epithelia:

- Simple squamous, like the endothelium lining the heart, blood vessels, and lymphatic vessels.
- Simple cuboidal, like the surface epithelium of the ovary.
- Simple columnar, some of which is ciliated. The lining of the small intestine is nonciliated simple columnar epithelium.
- Pseudostratified epithelium, which, as the name implies, appears to be stratified but actually contains only one layer of cells, since all are attached to the basement membrane. However, some cells do not reach the apical surface. The best known example is the ciliated pseudostratified columnar epithelium present in the respiratory passages.
- Stratified squamous keratinized, found mainly in the skin. Keratinized cells are dead cells that contain a large amount of the protein keratin.
- Stratified squamous nonkeratinized, which lines wet cavities such as the mouth, esophagus, and vagina.
- Stratified cuboidal, which lines the follicles in the ovaries and seminiferous tubules in the testes.

[viii] However, not all mesenchyme is derived from mesoderm.

[ix] The terms "basal lamina" and "basement membrane" were often used interchangeably until electron microscopy revealed that the basal lamina is one of three layers that form the basement membrane.

- Stratified columnar, a rare type found in limited regions of the pharynx, larynx, anal canal, and male urethra.
- Transitional, which lines the urinary bladder, the ureters, and the upper part of the urethra. This tissue is so-named because cells in its surface layer change shape as the bladder alternately distends to a larger size and then collapses to a smaller size.

Epithelia are functionally classified into two main groups: covering epithelia and glandular epithelia. Covering epithelia are those that cover body surfaces exposed to the outside. Most glands are composed primarily of epithelium whose cells are specialized to produce a fluid secretion that differs in composition from blood or extracellular fluid. However, this is an arbitrary division because the categories overlap – for example, the surface epithelium of the stomach is a covering epithelia in which **all** cells secrete mucus. In fact, glands always develop from covering epithelia by means of cell proliferation and invasion of the underlying connective tissue, followed by further differentiation.

The epithelia that form the glands of the body can be classified into unicellular glands, consisting of isolated cells, and multicellular glands, composed of clusters of cells. For example, the goblet cell, found in the lining of the small intestine or of the respiratory tract, is a unicellular gland. The pancreas is an example of a multicellular gland.

Glands can also be classified as exocrine or endocrine. Exocrine glands secrete substances that are delivered to the surface of the body because they retain their connection with the surface epithelium, in the form of ducts lined with epithelial cells, from which they developed. Examples of these glands are sudoriferous (sweat) glands, goblet cells, and salivary glands. Secretory products include mucus, sweat, oil, saliva, wax, milk, and digestive enzymes. Endocrine glands secrete substances into the surrounding tissue fluid because their connection with the surface epithelium from which they arose was obliterated during development. Their secretions eventually are picked up and transported to their site of action by the bloodstream. Examples of these glands are the thyroid, adrenal, and pituitary. Secretory products include hormones and plasma proteins.

Some organs have both exocrine and endocrine functions. In some of these organs, one cell type may function both ways. For example, hepatocytes secrete bile into the duct system of the liver and also produce secretions that enter the bloodstream. In other organs, some cells are specialized in exocrine secretion and others are concerned exclusively with endocrine secretion. For example, the acinar cells of the pancreas secrete digestive enzymes into the intestines while the islet cells secrete hormones that enter the blood.

TOPIC 8.2. CONNECTIVE TISSUE

Connective tissue is the most diverse, abundant, and widely distributed tissue in the human body. Connective tissues are not located on free (unattached) surfaces. All except cartilage have a nerve supply, and all except cartilage and tendons have a rich blood supply. The functions of connective tissue include:

- Binding together, supporting, and strengthening other tissues. A tendon attaches a skeletal muscle to bone, another muscle, or the dermis of the skin, and a ligament binds bone to bone. Bones support the body, and cartilage supports the ears, nose, trachea, and bronchi.
- Protecting and insulating the internal organs. The cranium, ribs, and sternum protect delicate organs such as the brain, heart, and lungs, while adipose tissue below the skin thermally insulates the body.
- Compartmentalizing certain structures, such as skeletal muscles.
- Transporting substances. Blood and lymph transport gases, nutrients, wastes, hormones, and blood cells.
- Storing energy and minerals. The fat in adipose tissue can be released and the calcium and phosphorus in bone can be drawn from, when needed.

One general characteristic of connective tissue is that its cells are widely separated by the extracellular matrix, which determines the tissue's properties. Matrix is usually secreted by the connective tissue cells and adjacent cells and may be fluid, gel, or solid. It consists of ground substance and fibers. The ground substance fills the space between and helps bind together the cells and fibers of connective tissue. It is colorless, transparent, and homogenous. Three types of fibers are found in connective tissue: collagen, reticular fibers, and elastic fibers.

Collagen is actually a group of related structural proteins that evolved several hundred million years ago, modified by environmental influences and the functional requirements of the animal body to varying degrees of rigidity, elasticity, and strength. Collagen is the most abundant protein in the human body, constituting about 25 percent of the body's total protein. A collagen fibril consists of collagen molecules packed into an organized overlapping bundle. Collagen fibers are bundles of fibrils. Although more than two dozen types of collagen have been described, over 90 percent of the collagen in the body are types I, II, III, and IV. Collagen type I is the most abundant, is the type that occurs as collagen fibers, and is the main component of bone. Collagen type II is the main component of cartilage and the vitreous humor of the eye. It occurs as very thin fibrils. Collagen type III is usually associated with collagen type I in the tissues and is the main component of reticular fibers. Reticular fibers are made of thin fibers of collagen type III coated with glycoprotein. They are found especially in reticular connective tissue, where they branch extensively. Collagen type IV is present in all basement membranes and in the lens of the eye. It forms sheets instead of either fibrils or fibers. Elastic fibers are made of a protein called elastin, which stretches and recoils like a rubber band. The presence of these fibers in the skin, lungs, and arteries allows these structures to recoil after they have been stretched.

Connective tissues can be divided into four major subclasses:

- Embryonic connective tissue known as mesenchyme, mentioned earlier.
- Fibrous connective tissue, also known as fibroconnective tissue.

- Supportive connective tissue.
- Fluid connective tissue.

Fibrous connective tissue is the most diverse and widely distributed type of permanent connective tissue in the body. Fibers are especially prominent. This connective tissue contains fibroblasts, which secrete the matrix. Some types of fibrous connective tissue contain adipocytes (fat cells), which store triglyceride (fat). There are two broad categories of fibroconnective tissue: loose connective tissue and dense connective tissue.

In loose connective tissue, the fibers are loosely woven and ground substance occupies much of the space. There are three types: areolar connective tissue, adipose connective tissue, and reticular connective tissue. Areolar connective tissue consists of all three types of fibers, several cell types, and a semifluid ground substance. Nearly every epithelium in the body rests on a layer of areolar tissue and depends on its blood vessels for nutrition and waste removal. Adipose tissue is dominated by adipocytes. The number of adipocytes and the quantity of stored triglyceride are quite stable in a person, but there is a constant turnover of stored triglyceride. The space between adipocytes is occupied by areolar tissue, reticular tissue, and blood capillaries. Reticular connective tissue is a mesh of reticular fibers and reticular cells (fibroblasts) that forms the structural framework (stroma) of organs and tissues such as the liver, spleen, lymph nodes, and bone marrow. The space between the fibers is occupied by blood cells.

In dense connective tissue, the fibers are more numerous, thicker, and closely packed, and the cells and ground substance occupy relatively little space. There are three types: dense regular connective tissue, elastic connective tissue, and dense irregular connective tissue. Dense regular connective tissue has bundles of parallel collagen fibers, all running in the same direction, and a few fibroblasts. The arrangement of fibers reflects the use of this tissue in tendons and ligaments, which pull in fairly predictable directions. Since dense regular connective tissue contains few blood vessels, tendons and ligaments are slow to heal. Elastic connective tissue has branching elastic fibers, closely packed collagen fibers, and a greater number of fibroblasts than dense regular connective tissue. It is found in the walls of large and medium arteries and in the lungs, trachea, and bronchial tubes.[x] Dense irregular connective tissue is similar to dense regular, but has randomly arranged bundles of collagen fibers that allow it to resist stresses that may be applied in unpredictable directions. It occupies most of the dermis of the skin.

Supportive connective tissues have a stiff or hard matrix and provide support for the body or its parts. There are three broad categories of supportive connective tissue: cartilage, bone tissue, and dental tissues.

Cartilage has a matrix that is rubbery due to the presence of chondroitin sulfate and that contains collagen fibers and sometimes elastic fibers. Cells called chondroblasts produce the matrix and then become trapped in tiny cavities (lacunae), whereupon they are called chondrocytes. Because cartilage is avascular, its cells must obtain nutrients and get rid of wastes by diffusion through the stiff matrix. Three types of cartilage have evolved, each

---

[x]   Elastic connective tissue is sometimes considered a type of dense regular connective tissue.

exhibiting variations in matrix composition: hyaline, elastic, and fibrocartilage. Hyaline cartilage is the most abundant but weakest cartilage in the body. Its matrix contains primarily type II collagen and a resilient gel as its ground substance. It is found in the nose, at the ends of the long bones and ribs, and in the supporting rings of the trachea. Elastic cartilage has a matrix containing many fine elastic fibers as well as collagen type II fibrils. It is more flexible than hyaline cartilage and is found, for example, in the external ear. Fibrocartilage is the strongest cartilage in the body. Its matrix contains a great number of collagen type I fibers. Fibrocartilage absorbs shock and reduces friction at joints and is found, for example, in the discs between vertebrae.

Bone (osseous) tissue has a rigid matrix composed of mineral salts and collagen fibers. Cells called osteoblasts produce the matrix and then become trapped in lacunae, whereupon they are called osteocytes. There are two types of bone tissue: compact and spongy. Compact bone tissue is organized into osteons (Haversian systems) that contain lamellae, lacunae, osteocytes, canaliculi (singular: canaliculus), and central (Haversian) canals. It forms the external surface of all bones. Spongy (cancellous) bone tissue consists of thin columns (trabeculae), the spaces between which are filled with bone marrow. It is found at the ends of the long bones and in the middle of nearly all the others. It is always enclosed by more durable compact bone tissue.

Table 4. Types of Connective Tissue

Fibrous connective tissue	**Loose CT**	**Areolar (loose) CT**	Collagen, reticular, and elastin fibers
		Adipose tissue	Cells store triglyceride; very little matrix
		Reticular CT	Many reticular fibers
	Dense CT	**Regular**	Bundles of parallel collagen fibers
		Irregular	Bundles of nonparallel collagen fibers
		Elastic CT	Many elastic fibers
Supportive connective tissue	**Cartilage**	**Hyaline cartilage**	Fine collagen fibers
		Elastic cartilage	Many elastin fibers
		Fibrocartilage	Strong collagen fibers
	Bone tissue	**Compact bone**	Lamellae arranged in osteons
		Spongy bone	Lamellae arranged in trabeculae
	Dental tissue	**Dentin**	Gives tooth shape and rigidity
		Cementum	Helps bind tooth to jaw
Fluid connective tissue	**Blood tissue**		Found in cardiovascular system; matrix (plasma) is liquid
	Lymph		Same composition as tissue fluid, but found in lymphatic system

Dental tissues are bone-like calcified connective tissues that form the teeth. However, teeth are **not** considered to be bones and thus are **not** part of the skeletal system. There are two types of dental tissues: dentin and cementum. Dentin, which is harder than bone, forms most of a tooth, giving it shape and rigidity. It underlies and supports a thinner layer of harder enamel that covers the upper portion of the tooth. However, enamel is not a connective tissue but a noncellular secretion. Cementum and other tissues bind the root of a tooth to its socket in the jaw.

Fluid connective tissues travel through tubular vessels. They are also referred to as transport tissues, reflecting their primary function of transporting cells and dissolved substances within the body. There are two principal types of fluid connective tissue: blood tissue and lymph.

Blood tissue (or simply blood) contains a liquid matrix called blood plasma and cells or cell derivatives called formed elements. The formed elements, which do not secrete the matrix, consist of the following: erythrocytes, also known as red blood cells or RBCs, which transport oxygen and carbon dioxide; leukocytes, also known as white blood cells or WBCs, which are involved in phagocytosis, immunity, and allergic reactions; and platelets, which function in blood clotting.

Lymph is interstitial (extracellular) fluid that flows in lymphatic vessels. It consists of a clear fluid similar to blood plasma but with much less protein. The composition of lymph varies from one part of the body to another.

A membrane in the body is a thin, sheet-like layer of tissue that covers a surface, lines a cavity, or divides a space or organ. There are two principal types of body membranes: epithelial and synovial.

An epithelial membrane consists of an epithelial layer and an underlying layer of connective tissue. There are several types. Mucous membranes (mucosae) line body cavities that are open to the outside, such as the gastrointestinal tract, respiratory tract, and urinary tract. Serous membranes (serosae) line closed cavities that contain internal organs. The membrane that lines the wall of the cavity is called the parietal layer and the membrane that covers the organs is called the visceral layer. Both layers are continuous. The simple squamous epithelium, known as mesothelium, of the membrane secretes a lubricating fluid into the space between the layers. This serous fluid reduces friction between the organs and the walls of the cavity that surround them. The cutaneous membrane is another term for the skin.

A synovial membrane consists of areolar connective tissue with synoviocytes, some of which resemble macrophages and others fibroblasts. Synovial membranes lack any epithelium and line (synovial) cavities that surround articulating bones. The synoviocytes secrete a lubricating synovial fluid that reduces friction between the bones during movements.

TOPIC 8.3. MUSCLE TISSUE

Muscle tissue consists of elongated cells that are specialized to generate tension. The functions of this tissue include providing motion, maintaining posture, and producing heat. There are three types of muscle tissue: skeletal, cardiac, and smooth.

Skeletal muscle tissue functions to move bones (and in some cases, skin and other soft tissues, such as the tongue and lips) and is voluntary, that is, under conscious control. It consists of long, cylindrical, parallel cells, usually referred to as fibers, that are striated, meaning that alternating light and dark bands, called striations, are visible when the cells are stained and examined microscopically. The muscle fiber is so filled with contractile proteins that its multiple nuclei are located adjacent to the plasma membrane. The term "skeletal muscle" refers to both this type of tissue as well as the muscles which it forms – organs that also contain connective tissue, blood vessels, and nervous tissue.

Cardiac muscle tissue forms most of the heart wall and consists of striated cells. It is involuntary and contracts without nervous stimulation. In fact, isolated cardiac muscle cells are autorhythmic – they produce action potentials spontaneously at regular time intervals! Unlike the fibers of skeletal muscle tissue, cardiac muscle cells are much shorter and contain more cytoplasm and more mitochondria. They are branched, contain only one centrally located nucleus, and are also distinguished by structures called intercalated discs that join them end-to-end.

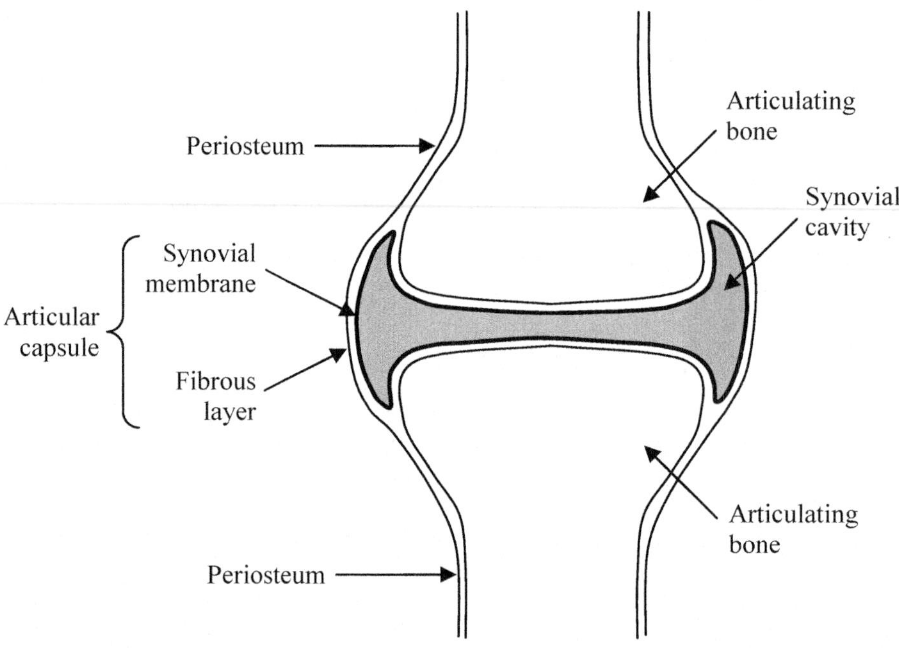

9. *Schematic diagram showing the position of a typical synovial membrane within an articular capsule.*

Smooth muscle tissue generally occurs in the walls of hollow internal structures (blood vessels and viscera) and is usually involuntary. Its cells can contract without nervous stimulation; some are even autorhythmic. Smooth muscle cells are relatively short and fusiform (narrow with two tapering ends), contain a single, oval centrally located nucleus, and are nonstriated (lack striations).

TOPIC 8.4. NERVOUS TISSUE

Nervous tissue is composed of neurons (nerve cells) and more numerous cells called neuroglia or glial cells that protect and support the neurons. It has scarcely any extracellular matrix. Neurons detect stimuli, convert stimuli into nerve impulses, and conduct these signals to other neurons, muscle cells, or glands. Most of the cytoplasm, including the nucleus and most other organelles, of a neuron are located in its cell body (soma). Two types of processes, called neuron fibers, typically extend from the soma: the dendrite and the axon. There are usually multiple dendrites, which by definition conduct nerve impulses to the cell body. Dendrites are typically short and highly branched. There is usually only one axon, which by definition conducts nerve impulses away from the cell body. It is typically much longer than the dendrites and has relatively few branches, except at its distal end.

TOPIC 9. THE INTEGUMENTARY SYSTEM

The integumentary system consists of the skin and its accessory organs: hair, nails, and cutaneous glands. The skin is the largest organ in the body. In an adult, it covers an area of about 1.5 to 2.0 square meters and makes up about 15 percent of the body by weight. The functions of the integumentary system include:

- Helping to control body temperature (thermoregulation).
- Preventing excessive loss or gain of water.
- Protecting the body from infections, physical injury, solar radiation, and chemicals.
- Detecting sensations, such as touch, hot and cold, and pain.
- Performing a step in the synthesis of vitamin D.
- Participating in excretion of certain metabolic wastes. A waste is any substance of no use to the body or present in excess of the body's needs. More specifically, a metabolic waste is a waste substance produced by the metabolism, such as carbon dioxide in the tissues. (Thus, the undigested food in feces is a waste but not a metabolic waste.) Excretion is the process of eliminating metabolic wastes from a cell or from the body. The integumentary system excretes water, inorganic salts, lactic acid, and urea in the sweat.[xi]

The skin consists of two major layers: the outer, thinner epidermis and the inner, thicker dermis. The epidermis is composed of keratinized stratified squamous epithelium and contains four major types of cells. Ninety percent of all cells in the epidermis are keratinocytes, which produce keratin, a durable and waterproof protein. Melanocytes produce melanin, a pigment. Langerhans cells help in the immune response. Merkel cells are thought to function in the sensation of touch. The epidermis can be divided into five zones, described as follows, from deep to superficial. The stratum basale is a single

---

[xi]   Other systems that participate in excretion are: the respiratory system, which excretes carbon dioxide, small amounts of other gases, and water; the digestive system, which excretes water, salts, carbon dioxide, lipids, bile pigments, cholesterol, and other metabolic wastes; and the urinary system, which excretes a broad range of metabolic wastes.

layer of cuboidal to low columnar cells resting on the basement membrane of the epithelium. Its cells continuously divide to produce all the other layers. The stratum spinosum consists of several layers of keratinocytes, so-named for its "spiny" appearance that is the result of the fixation process prior to microscopic examination. The stratum granulosum consists of two to five layers of flat cells containing coarse, dark-staining granules that give this layer its name. The stratum lucidum is a thin translucent zone seen only in thick skin, which is found on the palms, the anterior surfaces of the fingers, and the soles of the feet. The stratum corneum consists of 25 to 30 layers of flat, dead, keratinized cells that are continuously shed and replaced by cells from the deeper strata. This layer is thicker than average in thick skin. It is thinner than average in thin skin, which is found in especially sensitive areas of the body, such as the lips, eyelids, eardrums, and some parts of the genitals. The accumulation of the waterproofing and protective keratin, known as keratinization, occurs as cells move from the deepest layer of the epidermis to the surface layer.

The dermis is fibrous connective tissue composed mainly of collagen with some elastic and reticular fibers. It is divided into two zones: the papillary region and the reticular region. The papillary region is the upper 20 percent of the dermis, which consists of areolar connective tissue. It is distinguished by dermal papillae, which are folds of dermis that project into the epidermis. This arrangement resists slippage of the epidermis across the dermis when a shearing force is applied to the skin. These projections can also produce ridges in the overlying epidermis that increase friction and thus provide a better gripping surface. Tactile (Meissner's) corpuscles or corpuscles of touch, which are sensory receptors for touch, are also found in the papillary region. The reticular region is the lower 80 percent of the dermis, which consists of dense irregular connective tissue and contains nerves, blood vessels, and small clusters of adipocytes. Sudoriferous (sweat) glands and hair follicles originate here and in the deeper hypodermis. Associated with the hair follicles are the arrector pili muscle and sebaceous (oil) glands.

The layer below the skin, called the hypodermis, is also known as the subcutaneous layer or superficial fascia. Composed of areolar and adipose tissues, it anchors the dermis to underlying tissues and organs, stores fat, and provides thermal insulation. Lamellated (Pacinian) corpuscles, which are sensitive to deep pressure, are also found in the hypodermis.

The color of skin is due to melanin, carotene, and hemoglobin. Melanin is produced by melanocytes in the stratum basale. It produces a variety of brown, black, tan, yellowish, and reddish hues. People of all races have about equal numbers of melanocytes; the difference in their pigmentation is due to differences in the amount and dispersal of melanin produced. Carotene is found mostly in the dermis. It is a yellow pigment not synthesized by the body, but acquired from egg yolks and yellow and orange vegetables in the diet. Hemoglobin is found in erythrocytes that flow through dermal capillaries. It imparts pink and flesh tones to the skin.

Sudoriferous glands, also called sweat glands, are the most numerous cutaneous glands and consist of two types: eccrine and apocrine. Eccrine sweat glands produce secretions that exit through a duct leading to a sweat pore on the skin surface. These are active throughout life and function primarily to cool the body by evaporation of the sweat they produce. Apocrine sweat glands produce secretions that exit through a duct that opens into adjacent hair follicles. They are not active until puberty, and then respond especially to stress and sexual stimulation. Apocrine sweat glands are localized in the same areas of the body as the pubic, axillary, and male facial hair, suggesting that these glands are similar to other mammalian scent glands in function.

Hairs, or pili (singular: pilus), and nails are composed of dead, keratinized epidermal cells. However, the skin contains pliable soft keratin, while the hair and nails contain mostly hard keratin. Hard keratin is tougher and more compact. Hair is found almost everywhere on the body except the lips, nipples, parts of the genitals, and certain areas of the hands and feet. Most hair of the human trunk and extremities is probably best interpreted as vestigial, with little present function. Each hair consists of a shaft, root, and bulb. The shaft projects above the surface of the skin and the root penetrates into the dermis. The follicle is a diagonal tube that surrounds the root. The bulb is a swelling at the deepest portion of the follicle where the hair originates. Except near the bulb, all the tissue of a hair is dead. Hair grows due to cell division in the stratum basale within the bulb of the hair follicle. Hair color is primarily due to the amount of melanin produced by melanocytes in the stratum basale.

Associated with each hair follicle is a hair root plexus, sebaceous gland, and arrector pili muscle. The hair root plexus consists of dendrites of neurons. The dendrites entwine each follicle and are sensitive to hair movements. The sebaceous gland produces an oily secretion called sebum that is discharged into a hair follicle or directly onto the skin surface. Sebum prevents dehydration of hair and skin, and inhibits growth of certain bacteria. If enlarged, sebaceous glands may develop into blackheads, and sometimes into pimples and boils. The arrector pili muscle consists of a bundle of smooth muscle cells which extends from dermal collagen fibers to the follicle. Its contraction causes elevation of the hair shaft, known as piloerection.

Nails are located over the dorsal surfaces of the distal portions of the digits. Unlike the claws of most mammals, flat nails permit more sensitive fingertips and help to grasp and manipulate objects. Each nail consists of a nail plate and a nail matrix. The nail plate, in turn, consists of three parts: the body is the visible part of the nail plate (pink in color); the free edge is the part that extends past the distal end of the digit (white); and the root is the part that is buried in a fold of skin. The nail matrix is located deep to the proximal end of the nail root. New cells are added to the nail plate due to cell division in the stratum basale within the nail matrix.

TOPIC 10. BONE TISSUE AND BONES

A bone is an organ that is made up of several different tissues besides bone (osseous) tissue, including cartilage, dense connective tissues, epithelium, various blood-forming tissues, adipose tissue, and nervous tissue. Bones may be classified on the basis of shape into five groups:

- Long bones, such as the femur or humerus, have greater length than width. They serve as rigid levers that are acted upon by skeletal muscles to produce body movements.
- Short bones, such as the carpal (wrist) and tarsal (ankle) bones, tend to be cuboidal, that is, their length and width are about equal. They have limited motion and glide across one another.
- Flat bones, such as the scapula and sternum, are platelike and have broad surfaces for muscle attachment. They enclose and protect soft organs.
- Irregular bones, such as the vertebrae and the calcaneus (heel bone), have an odd, complex shape.
- Round (sesamoid) bones are circular. They are usually pea-sized and found in certain tendons of the hands and feet; the largest is the patella.

Sutural (Wormian) bones are a category of bone based on their location rather than their shape. They are small bones located within the sutures (joints) between certain cranial bones and vary greatly in number from person to person.

All bones are enclosed by a layer of thick, dense, white osseous tissue called compact, or dense, bone. A typical long bone has several notable features:

- The diaphysis is the shaft, or body, of a long bone. It encloses the medullary cavity, a space that contains yellow bone marrow (rich in adipocytes) in adults.
- The epiphyses (singular: epiphysis) are the distal and proximal ends of a long bone. The central spaces of the epiphyses of long bones and the middle of nearly all others contain the other type of osseous tissue called spongy, or cancellous, bone. Spongy bone is more delicate than compact bone and is named for its sponge-like appearance.
- The metaphyses are the regions in a mature long bone where its diaphysis joins the epiphyses. In a growing bone, each metaphysis contains an epiphyseal plate, which is a layer of hyaline cartilage that is responsible for the longitudinal growth of the long bone. When lengthwise growth stops, the epiphyseal plate is replaced by bone and is called the epiphyseal line.
- The articular cartilage is a thin layer of hyaline cartilage that covers the surface of each end of the bone where it forms a joint (articulation) with another bone. It reduces friction between the articulating bones.
- The periosteum, a layer of dense irregular connective tissue and associated cells, covers the external bone surface wherever it is not covered by the articular cartilage. It enables bone to grow in width, but not in length.
- The endosteum, a layer of reticular connective tissue and associated cells, lines all internal surfaces of cavities within the bone.

Bone tissue is composed of an abundant extracellular matrix that surrounds widely separated cells. There are four principal types of bone cells: osteogenic cells, osteoblasts, osteocytes, and osteoclasts. Osteogenic, or osteoprogenitor, cells are stem cells derived from mesenchyme. They are found in periosteum, in endosteum, and in the canals within bone that contain blood vessels. They are the only bone cells to undergo cell division, producing osteoblasts. Osteoblasts are cells that secrete the extracellular matrix of bone. They are incapable of cell division but large numbers may be produced by osteogenic cells under the influence of stress or fractures. Osteocytes are osteoblasts that become trapped within tiny spaces (lacunae) as they deposit extracellular matrix. Osteocytes no longer produce matrix but they maintain bone tissue. Osteoclasts are very large multinuclear cells that dissolve bone. They form by the fusion of monocytes, one of the five kinds of white blood cells. They are found in endosteum.

The extracellular matrix of bone tissue is composed of abundant mineral salts, mostly hydroxyapatite (calcium phosphate) and collagen fibers. Crystallized mineral salts give bone tissue hardness and collagen fibers give it flexibility, a combination of strength and resilience similar to that of reinforced concrete. The matrix of both mature compact and mature spongy bone tissue is organized in layers called lamellae.

The basic structural unit of compact bone tissue is the osteon (Haversian system). An osteon consists of a central (Haversian) canal containing blood and lymph vessels and nerves, surrounded by concentrically arranged lamellae. Between the lamellae are the small almond-shaped lacunae occupied by osteocytes in living bone. Delicate channels called canaliculi connect each lacuna to its neighbors and also connect the innermost lacunae to the central canal. Cytoplasmic processes of cells in adjacent lacunae, which extend into and meet within the canaliculi, are connected by gap junctions. In this way, nutrients pass from blood vessels in the central canal through canaliculi to the osteocytes in the nearest lacunae, which then pass them on through canaliculi to more remote ones; wastes pass in the opposite direction.

The lamellae of spongy bone tissue are organized to form thin columns, or trabeculae. The interconnecting spaces between the trabeculae are occupied by bone marrow. There are few, if any, osteons in spongy bone because each osteocyte is close enough to the blood vessels in the marrow to obtain nutrients and eliminate wastes directly by diffusion.

Marrow is the soft tissue that occupies the medullary cavities of long bones, the spaces between the trabeculae of spongy bone, and the larger Haversian canals. There are two types: red and yellow. Red bone marrow (myeloid tissue) produces blood cells. They develop within a mesh of reticular connective tissue that also contains scattered adipocytes. Fetuses and children have only this type of marrow, found in almost every bone of their skeletons. Yellow bone marrow, composed mainly of adipocytes, does not produce blood cells. In adults, yellow bone marrow occupies most of the long bones and red bone marrow is restricted to the axial skeleton and proximal heads of the humerus and femur.

Surface markings are structural features visible on the surfaces of bones. Each marking has a specific function. There are two broad categories of surface markings: 1) depressions and openings and 2) processes, which are projections or outgrowths on bone.

Depressions and openings allow the passage of soft tissues, such as nerves, blood vessels, ligaments, and tendons. They include: the foramen (plural: foramina), which is a rounded opening through a bone (for example, the mandibular, or mental, foramen of the mandible); the fossa, which is a flattened or shallow surface (for example, the glenoid fossa of the scapula); the meatus, which is a tubelike passageway through a bone (for example, the external auditory meatus of the temporal bone); and the sinus, which is a cavity or hollow space in a bone.

Forming joints is the function of some processes, such as: the condyle, which is a large, rounded, articulating knob (for example, the lateral condyle of the femur); the facet, which is a smooth flat articular surface (for example, the superior articular facet of a vertebra); and the head, which is a prominent, rounded, articulating proximal end of a bone (for example, the head of the femur). Forming attachment points for connective tissue is the function of other processes, such as: the crest, which is a narrow, ridgelike projection (for example, the iliac crest of the os coxa); the spine, which is a sharp, slender process (for example, the spine of the scapula); the trochanter, which is a massive process found only on the femur; the tubercle, which is a small, rounded process (for example, the greater tubercle of the humerus); and the tuberosity, which is a large, roughened process (for example, the ischial tuberosity of the os coxa).

The formation of bones is called ossification, or osteogenesis. It begins during the sixth or seventh week of embryonic development. Mesenchyme (embryonic connective tissue) forms a bone in one of two ways. In endochondral ossification, mesenchymal cells become chondroblasts that produce a "model" of the bone composed of hyaline cartilage. Then, osteoblasts gradually replace the cartilage with bone. Most bones of the body, except the cranial bones, form this way. In intramembranous ossification, bones form directly from mesenchyme without first going through a cartilage stage. This occurs in the cranial bones of the skull. The clavicles, mandible, and facial bones form by a combination of the two methods. During childhood, bones throughout the body grow in thickness as periosteal osteoblasts add new bone tissue to the external surfaces of bones (appositional growth), and long bones grow in length as cartilage is replaced by osseous tissue on the diaphyseal side of the epiphyseal plates. Ossification of most bones is usually completed by age 25.

Even after a bone is fully formed, it remains a metabolically active organ, maintaining itself and exchanging minerals with the tissue fluid. Remodeling is the ongoing replacement of old osseous tissue by new osseous tissue. It involves bone resorption and bone deposition. Resorption is the process of dissolving the extracellular matrix and releasing its minerals into the body. It is performed by osteoclasts. Deposition, also known as mineralization or calcification, is the process of adding crystallized mineral salts to the framework of collagen fibers in the extracellular matrix, thereby hardening the tissue. This process is performed by osteoblasts. Remodeling provides several advantages. First, new bone tissue is less brittle and stronger than old bone. Secondly, bone can be altered for proper support based on mechanical stresses as remodeling occurs. The main mechanical stresses on bone result from the contraction of skeletal muscles and the pull of gravity. Mechanical stress increases bone strength by increasing production of collagen fibers and deposition of mineral salts. Removal of mechanical stress weakens bone through resorption. Finally, remodeling also replaces injured bone tissue with new tissue.

A fracture is any break in a bone due to sudden, excessive mechanical stress. Types of fractures include:

- Closed (formerly called simple), in which the broken bone does not pierce the skin.
- Open (formerly called compound), in which the broken ends of bone pierce the skin.
- Complete, in which the bone is broken into two parts.
- Partial, in which the bone is broken longitudinally but not separated into two parts.
- Greenstick, in which a break on the outer arc of the bone is incomplete (occurs only in children).
- Impacted, in which one bone fragment is driven into the medullary space or spongy bone of the other.
- Comminuted, in which the bone breaks into three or more pieces.
- Spiral, in which the break is ragged due to twisting of bone.
- Hairline, a fine crack in which sections of bone remain aligned. It is common in the skull.

The repair of an uncomplicated fracture takes place in about 8 to 12 weeks, while complex fractures take longer. The healing process can be divided into four stages.

- First, a fracture hematoma (mass of clotted blood) forms, usually 6 to 8 hours after the injury. It forms in the space between the broken bones due to ruptured blood vessels, and the area becomes inflamed and swollen.
- Next, a fibrocartilaginous (soft) callus forms. In this stage, blood vessels grow into the hematoma, and fibroblasts and osteogenic cells migrate into the tissue. Fibroblasts deposit collagen and some osteogenic cells become chondroblasts, which produce patches of fibrocartilage.
- The third stage is the formation of a bony (hard) callus. Osteoblasts produce trabeculae of spongy bone, converting the fibrocartilaginous callus into a bony callus that acts like a splint to join the broken bones together. It takes about 4 to 6 weeks for a bony callus to form, during which time it is often necessary to immobilize a broken bone by traction or a cast to prevent refracture.
- The last stage is remodeling. Osteoblasts convert spongy bone to compact bone at the periphery. Osteoclasts resorb spongy bone to build a new medullary cavity.

TOPIC 11. THE SKELETAL SYSTEM: THE BONES

The skeletal system consists of bones, joints, tendons, and ligaments. Its functions include:

- Supporting soft tissues and providing attachment sites for the tendons of most skeletal muscles.

- Protecting many internal organs from injury.
- Storing and releasing minerals, especially calcium and phosphorus.
- Providing the site for blood cell production by red bone marrow.
- Storing fat, especially in yellow bone marrow.

There are two principal divisions in the adult skeleton: the axial skeleton and the appendicular skeleton. The axial skeleton consists of about 80 bones located around the longitudinal axis of the body and can be divided into four parts: skull bones; the auditory ossicles and hyoid bone; the vertebral column; and the thoracic cage. The appendicular skeleton consists of about 126 bones and can also be divided into four parts: the pectoral (shoulder) girdle; the upper limbs (upper extremities); the pelvic (hip) girdle; and the lower limbs (lower extremities).

The skull consists of 22 bones divided into the cranial bones and the facial bones. The eight cranial bones form the cranial cavity and consist of: the frontal bone; two parietal bones; two temporal bones; the occipital bone; the sphenoid bone; and the ethmoid bone. The occipital bone has a midsagittal projection on its posterior surface, called the external occipital protuberance; you may be able to feel this structure as a definite bump, the most prominent protrusion on the back of your head, just above your neck. It is the site of attachment of a large ligament which helps to support the head. Each temporal bone has a zygomatic process that articulates with a temporal process of a zygomatic bone, forming a zygomatic arch. The human temporal bone represents the fusion of four bones found in lower mammals: the squamosal bone, which constitutes the side of the head and articulates with the jawbone; the petrosal bone, which contains the inner ear; the mastoid bone, which is behind the ear; and the tympanic bone, which contains the external auditory meatus and supports the tympanic membrane. Fourteen facial bones form the face and consist of: two nasal bones; two maxillary bones; two zygomatic bones: the mandible; two lacrimal bones; two palatine bones; two inferior nasal conchae; and the vomer bone. The mandible represents the fusion of two bones found in lower mammals, called the dentary bones.

There are a number of important features of the skull. Sutures are the immovable joints located between the skull bones. In an infant, the spaces between the cranial bones of the skull are filled with membrane and are called fontanels. Fontanels enable the shape of the fetal skull to be modified as it passes through the vagina during childbirth. They also permit rapid growth of the brain during infancy and become ossified by age two. Paranasal sinuses are paired cavities named for the bones in which they occur: the maxillary, frontal, sphenoid, and ethmoid sinuses. They are connected with the nasal cavity and lined with mucous membranes that are continuous with the lining of the nasal cavity. The orbits, formed by several bones, are deep sockets that house the eyeballs and associated structures. Lastly, the skull bones contain numerous, especially conspicuous foramina that are passageways for blood vessels and nerves.

Seven bones are closely associated with the skull but are not considered part of it, namely, the hyoid bone and six auditory ossicles. The hyoid bone, located in the upper neck, is a **U**-shaped bone that does not articulate with any other. It supports the tongue and is an attachment site for several tongue, neck, and pharyngeal muscles. Each middle-ear (tympanic) cavity contains three auditory ossicles: the malleus, the incus, and the stapes. They transfer vibrations from the eardrum (tympanic membrane) to the inner ear.

The vertebral column, informally called the spine, or backbone, is not a single bone but a chain of about 33 vertebrae, most of which are separated by intervertebral discs of fibrocartilage. Its functions include: surrounding and protecting the spinal cord; supporting the head; and serving as an attachment site for ribs and back muscles. The vertebrae are divided into five regions:

- Cervical vertebrae designated C1 to C7. C1 is known as the atlas and C2 the axis.
- Thoracic vertebrae designated T1 to T12.
- Lumbar vertebrae designated L1 to L5.
- Sacral vertebrae designated S1 to S5. They begin to fuse between the ages of 16 and 18 to form the sacrum by age 26. The sacrum (meaning "sacred") is so named because it was once considered to be the seat of the soul. The sacral promontory is the anterior and superior border of the first sacral vertebra and projects anteriorly into the pelvic cavity.
- Coccygeal vertebrae designated Co1 to Co4. The four (sometimes five) vertebrae fuse to form the coccyx (tailbone) by the age of 20 to 30. Although the coccyx is a remnant of the ancestral reptilian tail, it is not entirely useless; it provides attachment for muscles of the pelvic floor. However, surgical removal of the coccyx has no discernible effect on health.[66] Therefore, it is considered to be a vestigial structure.

Intervertebral discs composed of fibrocartilage are located between adjacent vertebrae, from C2 to S1. There is no disc between the atlas and the axis. The discs form strong joints, permit various movements of the spine, and absorb vertical shock.

Several curves of the spine can normally be seen in a lateral view (but viewed posteriorly, the spine is normally straight). The four normal curves are the cervical, thoracic, lumbar, and sacral (pelvic) curvatures. The cervical and lumbar curves are anteriorly convex, that is, curved toward the dorsal side, and the thoracic and sacral curves are anteriorly concave, that is, curved toward the ventral side. They function to: increase the strength of the spine; help maintain balance in the upright position; diminish shocks that are transmitted to the spine, and through it to the head, by locomotion in the erect position; and help protect the spine from fracture. The entire spine at birth is anteriorly concave, as it is in monkeys, apes, and most other

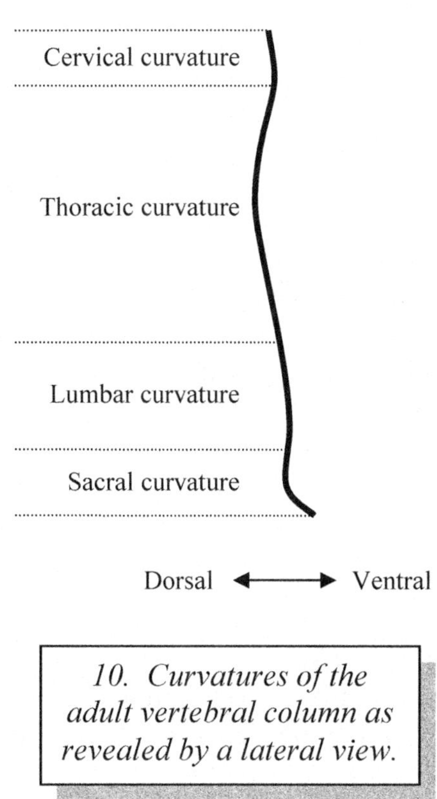

10. Curvatures of the adult vertebral column as revealed by a lateral view.

four-legged animals. However, the cervical curve develops as the infant begins to crawl and look forward and the lumbar curve develops as the toddler begins to walk.

The structural features of a typical vertebra are: the body, the vertebral arch, seven processes, and several foramina. The body is the thick, disc-shaped anterior portion that is the weight-bearing part. The vertebral arch consists of two laminae, which join to form the posterior portion, and two pedicles, which connect the laminae to the body. Seven processes arise from the vertebral arch: two transverse processes for muscle attachment; the spinous process for muscle attachment; two superior articular processes with facets to articulate with the vertebra above; and two inferior articular processes with facets to articulate with the vertebra below. The transverse processes of vertebrae T1 to T10 also have facets to articulate with the ribs. Each vertebra has a vertebral foramen located between the vertebral arch and body. All the vertebral foramina in the spine are aligned to form the vertebral (spinal) canal. An intervertebral foramen is located between the pedicles of adjacent vertebrae on each side. It permits the passage of a single spinal nerve.

The thoracic cage surrounds and protects organs in the thoracic cavity and upper abdominal cavity and provides support for the bones of the pectoral girdles and upper extremities. The bones of the thoracic cage are the sternum and ribs. The sternum, commonly called the breastbone, consists of three regions: the manubrium, body (gladiolus), and xiphoid process. There are twelve pairs of ribs (12 pairs), which can be divided into the true ribs (pairs 1 to 7) and the false ribs (pairs 8 to 12). The true ribs attach directly to the sternum by costal cartilage, whereas the false ribs are so called because their costal cartilages either fuse before attaching to the sternum (pairs 8 to 10) or do not attach to the sternum at all (pairs 11 and 12, also called the floating ribs).

The pectoral girdle attaches the bones of the upper extremities to the axial skeleton. It also functions as a point of origin for many muscles that move the upper limb, neck and trunk. The pectoral girdle consists of two clavicles and two scapulae. Each clavicle (collarbone) articulates medially with the manubrium of the sternum and laterally with the scapula. Each scapula (shoulder blade) articulates with the clavicle and also with the humerus. These are loose attachments that result in a shoulder far more flexible than that of most other mammals, but they also make the shoulder joint easy to dislocate.

Each upper extremity contains 30 bones. The humerus is the long bone of the upper arm, or brachium. Its head fits into the glenoid fossa of the scapula while its greater and lesser tubercles provide sites of attachment for muscles that move arm and shoulder. The radius and ulna are the bones of the forearm, or antebrachium. The radius is lateral to and shorter than the ulna, which is medial to and longer than the radius. The olecranon (elbow) is located on the posterior side of the proximal end of the ulna. Eight small bones arranged in two rows form the wrist, or carpus. The hand, or manus, contains 19 bones. The five metacarpal bones are located in the palm, or metacarpus. The distal ends of the metacarpals form knuckles when you clench your fist. The fourteen phalanges (singular: phalanx) are located in the digits. The pollex (thumb) has two phalanges while the other four digits each have three.

The pelvic (hip) girdle attaches the bones of the lower extremities to the axial skeleton. The pelvic girdle also provides a strong and stable support for the vertebral column and protects the urinary bladder, the internal reproductive organs, and a portion of the large intestine. The bones of the pelvic girdle consist of the two ossa coxae (also called innominate bones, hip bones, or coxal bones). Each os coxa in turn consists of an ilium, ischium, and pubis (pubic bone). The ilium is the largest and most superior part. It articulates posteriorly with the sacrum at a sacroiliac joint and has a conspicuous superior border called the iliac crest. The ischium is inferior and posterior to the ilium. It is significant for its thick, rough-surfaced ischial tuberosity, upon which you sit. The pubis is inferior and anterior to the ilium. It articulates with the other os coxae at the pubic symphysis. In a neonate (newborn), the ilium, ischium, and pubis are separate bones joined by cartilage, but by age 23 they fuse together. The acetabulum, a deep socket which receives the rounded head of the femur, is located where the ilium, ischium, and pubis meet together. The ischium and the pubis form a ring of bone enclosing the obturator foramen. Blood vessels and nerves pass anteriorly into each lower leg through this opening. The obturator is the largest foramen in the body, although it is nearly completely closed by a fibrous membrane (hence its name).

The bony pelvis consists of the two ossa coxae, the sacrum, and the coccyx. Thus, it is a combination of bones of the axial and appendicular skeletons. The flared parts of the ilium form the false (greater) pelvis, which encloses a space that is actually part of the abdominal cavity. The parts of the bony pelvis inferior to the false pelvis form the true (lesser) pelvis, which encloses a space called the pelvic cavity. The openings to the pelvic cavity are the upper pelvic inlet and lower pelvic outlet. The pelvic floor, or pelvic diaphragm, supports the pelvic organs, such as the urinary bladder, rectum, and uterus. It consists of two muscles, the levator ani muscle and the coccygeus muscle, and their fasciae. The perineum is the diamond-shaped region situated within the pelvic outlet, inferior to the pelvic floor and bordered by the pubic symphysis anteriorly, the ischial tuberosities laterally, and the coccyx posteriorly. The ischial tuberosities bisect the perineum into the anterior urogenital triangle and posterior anal triangle. The urogenital triangle contains the muscular urogenital diaphragm and the external genitalia while the anal triangle contains the anus. Both the sacroiliac joint and the pubic symphysis permit slight movement, especially during pregnancy when the ligaments of the latter are softened under the influence of hormones. The shape of the female pelvis is an adaptation for human pregnancy and childbirth (see Topic 23). There are a number of differences between female and male pelves. The female ilium is more flared than that of the male; therefore, the female has broader hips. The female pelvic inlet and outlet are wider. The female pelvic cavity is shallower, while the male pelvic cavity is more funnel-shaped. Female bones are lighter and thinner and the female pubic arch (angle at the pubic symphysis) is wider.

Each lower extremity contains 30 bones. The femur (thighbone) is the long bone of the thigh, or femoral region. It is the longest and strongest bone in the body. Its head fits into the acetabulum of the os coxae and its greater and lesser trochanters provide sites of attachment for the muscles of the legs and buttocks. The patella (kneecap) is a small, round bone located anterior to the knee joint. It forms within the tendon of the knee as a child begins to walk. The tibia and fibula are the bones of the lower leg, or crural region. The fibula is lateral to and more slender than the tibia while the tibia (shinbone) is medial to and larger than the fibula. Only the tibia bears the weight of the body. Seven tarsal bones are located in the ankle, or tarsus. The largest tarsal bone is the calcaneus (heel bone). The foot, or pes, contains 19 bones. The five metatarsal bones are located in the metatarsus, the intermediate region of the foot. The fourteen phalanges are located in the toes and resemble those of the hand both in number and arrangement. The big toe is called the hallux.

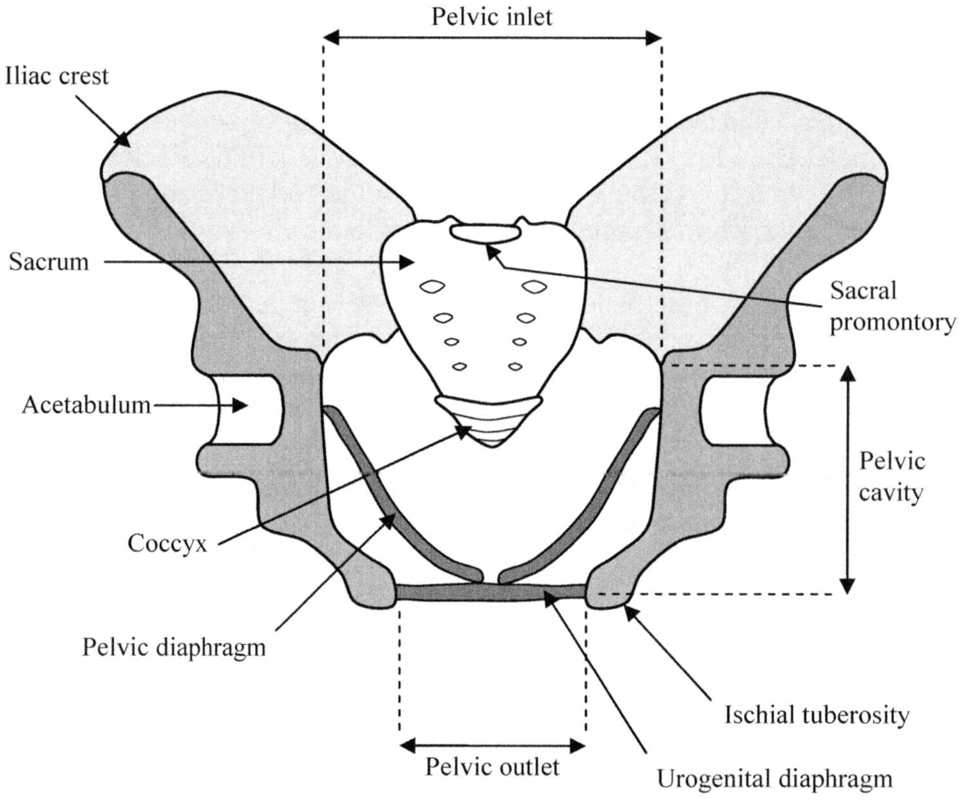

11. *Diagram of anterior view of frontal section of pelvis.*

TOPIC 12. THE SKELETAL SYSTEM: THE JOINTS

A joint, also called an articulation or arthrosis, is a point of contact between adjacent bones, between cartilage and bones, or between teeth and bones. The name of a joint is derived from the names of the articulating bones. For example, the atlanto-occipital joint is located between the occipital and the atlas, the humeroscapular joint is between the humerus and the scapula, and the sacroiliac joint is between the sacrum and the ilium of the os coxae. The structural characteristics of a specific joint affect the strength, magnitude of movement, and types of movement that may occur at that joint.

The structural classification of joints is based on the presence or absence of a joint (synovial) cavity and the type of connective tissue that binds the bones together. There are three types: fibrous joints, in which two bones are held together by collagen fibers extending from the matrix of one bone into the matrix of the other; cartilaginous joints, in which two bones are held together by cartilage; and synovial joints, in which the bones are separated by a cavity that contains a slippery lubricating fluid and are held together by a joint capsule.

The functional classification of joints is based on the magnitude of movement permitted. There are three types: the synarthrosis, which is an immovable joint; the amphiarthrosis, which is a slightly movable joint; and the diarthrosis, which is a freely movable joint. All diarthroses are synovial joints and all synovial joints are diarthroses.

A fibrous joint lacks a synovial cavity and the articulating bones are held very closely together by collagen fibers that emerge from one bone, cross the space between the bones, and penetrate into the matrix of the next, permitting little or no movement. There are three types: the suture, the gomphosis, and the syndesmosis.

In a suture, the collagen fibers holding the bones together are very short and allow for little movement. Therefore, it is functionally classified as a synarthrosis. Sutures are found only in the skull, for example, the coronal suture between the parietal and frontal bones. Sutures can be divided into three structural types: serrate, lap, and plane. The two bones of a serrate suture, such as the coronal suture, firmly interlock with each other along their serrated margins, like a dovetail wood joint. The two bones of a lap (squamous) suture have overlapping beveled edges, like a miter joint in carpentry. On the surface, the lap suture appears relatively nonserrated. The squamous suture between the temporal and parietal bones is an example. The two bones of a plane (butt) suture have straight, nonoverlapping edges. The bones merely border on each other, like two boards glued together in a butt joint. The joint between the palatine processes of the maxillary bones in the anterior roof of the oral cavity is an example.

In a gomphosis, a cone-shaped peg fits into a socket. Even though the teeth are not regarded as bones, the root of a tooth connected by a periodontal ligament to its socket (dental alveolus) in the mandible or the maxillary bone is classified as a gomphosis. The periodontal ligament consists of short collagen fibers that extend from the bone matrix of the jaw into the dental tissue, allowing for little movement.[xii] Therefore, a gomphosis is functionally classified as a synarthrosis.

In a syndesmosis, the collagen fibers holding the bones together are longer and the attached bones are more movable than in a suture or gomphosis. Therefore, a syndesmosis is functionally classified as an amphiarthrosis. The connecting collagen fibers are in the form of a ligament between the bones, such as the interosseous membrane between the diaphyses of the radius and the ulna and between the diaphyses of

---

[xii] In fact, the periodontal ligament and the periosteum of the alveolar bone are one and the same.

the tibia and the fibula.

Table 5. Examples of Joints and their Classification

Structural Type			Synarthrosis	Amphiarthrosis	Diarthrosis
Fibrous	Suture	Serrate	Coronal and sagittal sutures		
Fibrous	Suture	Lap	Squamous suture		
Fibrous	Suture	Plane	Palatine suture		
Fibrous	Gomphosis		Joint between tooth and socket		
Fibrous	Syndesmosis			Tibiofibular and radioulnar joints	
Cartilaginous	Synchondrosis			Epiphyseal plate and joints between ribs and sternum	
Cartilaginous	Symphysis			Intervertebral joints and pubic symphysis	
Synovial	Hinge				Elbow and knee joints
Synovial	Pivot				Atlantoaxial joint
Synovial	Ball-and-socket				Humeroscapular and coxal joints
Synovial	Saddle				Carpometacarpal joint of pollex
Synovial	Planar (gliding)				Intercarpal, intertarsal, and sternoclavicular joints
Synovial	Ellipsoidal				Radiocarpal and metacarpophalangeal joints

A cartilaginous joint lacks a synovial cavity and the articulating bones are tightly connected by cartilage, permitting little or no movement. There are two types: a synchondrosis and a symphysis. In a synchondrosis, the connecting tissue is hyaline cartilage. Examples include the epiphyseal plate of a growing long bone and the hyaline costal cartilage that joins a rib to the sternum. A synchondrosis is functionally classified as an amphiarthrosis. In a symphysis, the connecting tissue is a disc of fibrocartilage. Examples include the pubic symphysis, the joint between the manubrium and the body of the sternum, and the intervertebral joints. A symphysis is functionally classified as an amphiarthrosis. However, the collective effect of all 23 intervertebral discs gives the spine considerable flexibility.

The bones of a synovial joint are separated by a synovial (joint) cavity that contains a lubricant called synovial fluid (see Figure 9). A synovial joint is freely movable and therefore is functionally classified as a diarthrosis. It has an extensive nerve and blood supply. The surfaces of the articulating bones at the joints are covered with articular cartilage, consisting of hyaline cartilage that reduces friction between the bones during movement and helps to absorb shock. An articular capsule surrounds the synovial joint, encloses the synovial cavity, and unites the articulating bones. It has two layers: an outer fibrous capsule and an inner synovial membrane. The fibrous capsule usually consists of dense irregular connective tissue that is continuous with the periosteum of the articulating bones. The synovial membrane contains areolar tissue and secretes the synovial fluid into the synovial cavity.

Accessory structures of a synovial joint may include a tendon, ligament, bursa, or meniscus. A tendon is a strip or sheet of dense regular connective tissue that attaches a muscle to a bone, another muscle, or the dermis of the skin. A ligament is a strip or sheet of dense regular connective tissue that attaches one bone to another. Ligaments found within the joint capsule are called intracapsular, and those outside are called extracapsular; the knee joint, or tibiofemoral joint, has both types of ligaments. A bursa is a saclike structure containing synovial fluid that alleviates friction in joints such as the shoulder and knee joints. Tubelike bursae called tendon sheaths wrap around tendons where there is considerable friction, such as at the wrist and ankle. A meniscus, or articular disc, is a pad of fibrocartilage found between the articulating bones of the temporomandibular (TMJ), sternoclavicular, and tibiofemoral joints. The disc usually subdivides the synovial cavity into two separate spaces. The meniscus absorbs shock and pressure, guides the bones across each other, and reduces the chance of dislocation.

The structural classification of synovial joints is based on the shapes of the articulating surfaces.

- In a planar (gliding) joint, such as the intercarpal joints and the intertarsal joints, the articulating surfaces are flat or slightly curved. (Some anatomists consider these joints to be amphiarthroses.)
- In a hinge joint, such as the interphalangeal and tibiofemoral (knee) joints and the elbow joint, the convex surface of one bone fits into the concave surface of another bone. These joints are said to be monaxial because the articulating bones can move in only one (usually parasagittal) plane. The elbow joint actually consists of two joints that are enclosed in a single joint capsule: the humeroulnar joint, where the humerus articulates with the ulna, and the humeroradial joint, where the humerus articulates with the radius. The tibiofemoral joint is the largest and most complex joint of the body, actually consisting of three joints that share one synovial cavity: one tibiofemoral joint between the lateral condyles of the femur and the tibia; another tibiofemoral joint between the medial condyles of the femur and the tibia; and the patellofemoral joint between the patella and the

femur. The two tibiofemoral joints are modified hinge joints, while the patellofemoral is a planar joint.

- In a pivot joint, such as the atlanto-axial joint, the rounded or pointed surface of one bone fits into a ring-like ligament of another.
- In an ellipsoidal (condyloid) joint, such as the atlanto-occipital joint and the metacarpophalangeal joints, an elliptical (oval-shaped) projection of one bone fits into an elliptical depression in another bone. These joints are said to be biaxial because they can move in two directions, for example, flexing the index finger as if gesturing to someone, "come here," and moving the finger from side to side toward the thumb and away.
- The only saddle joint is the joint between the trapezium of the wrist and the metacarpal of the thumb. The articular surface of the metacarpal is saddle-shaped and the articular surface of the trapezium resembles the legs of a rider sitting in a saddle, making this joint more movable than an ellipsoidal or hinge joint. Thus, the trapeziometacarpal joint is responsible for the distinctive opposable thumb of primates.
- In a ball-and-socket joint, such as the humeroscapular, or glenohumeral, (shoulder) joint and coxal (hip) joint, the spherical surface of one bone fits into a cuplike depression of another bone. These joints are said to be multiaxial because the humerus or femur can be moved in more than two directions or planes.

The three basic types of movement that occur at synovial joints are rotational, gliding, and angular. In rotation, a bone revolves around its own longitudinal axis. For example, when the head is turned from side to side to signify "no," the atlas rotates on its longitudinal axis at the atlanto-axial joint. In a gliding movement, the nearly flat surfaces of bones move back and forth and from side to side without changing the angle between the bones. Gliding occurs at planar joints. In an angular movement, a change in the angle between bones occurs. There are various types of angular movement:

- Flexion decreases the angle, such as bending the head forward at the atlanto-occipital joint, as in signifying "yes."
- Extension increases the angle, usually to about 180°, such as straightening the elbow joint.
- Hyperextension increases the angle of a joint beyond 180°, such as looking up toward the ceiling, which hyperextends your neck.
- Abduction is movement of a body part away from the midline, such as moving the humerus laterally at the shoulder joint.
- Adduction is movement of a body part toward the midline, such as moving the humerus medially at the shoulder joint.
- Circumduction is movement of the distal end of a body part in a circle, such as moving the humerus in a circle at the shoulder.

Special movements occur only at certain joints:

- Elevation is an upward movement of a part of the body, such as elevating the mandible to close the mouth.

- Depression is a downward movement of a part of the body, such as depressing the mandible to open the mouth.
- Protraction is movement of a bone anteriorly (forward) in the transverse plane, such as jutting the jaw outward or hunching the shoulders forward.
- Retraction is a movement of a protracted part of the body back to the anatomical position.
- Inversion is turning the foot so that the sole is inward (facing medially).
- Eversion is turning the foot so that the sole is outward (facing laterally).
- Dorsiflexion is flexion of the foot upward, as when you stand on your heels.
- Plantar flexion is flexion of the foot downward, as when you stand on your toes.
- Supination is the rotation of the forearm so that the palm is upward/forward.
- Pronation is the rotation of the forearm so that the palm is downward/backward.
- Opposition is the movement of a thumb across the palm to touch the fingertips on the same hand.
- Reposition is movement of the opposed thumb back to the anatomical position.

TOPIC 13. MUSCLE HISTOLOGY

Although bones and joints provide leverage and form the framework of the body, motion results from alternating contraction (shortening) and relaxation of muscles. A muscle is an organ composed primarily of muscle tissue but also containing connective tissue, blood vessels, and other tissues. The three types of muscle tissue were described in Topic 8.3. Regardless of type, all muscle tissue has four key functions:

- Producing body movements, such as walking, running, and grasping.
- Stabilizing body positions, for example, standing, sitting, and holding the head upright.
- Storing and moving substances within the body, such as pumping blood, moving lymph, and peristalsis in the digestive tract and urinary tract.
- Producing heat that is used to maintain body temperature.

In addition, muscle tissue has four special properties:

- Electrical excitability, also possessed by neurons, is the ability to respond to certain stimuli by producing rapid changes in membrane potentials, called action potentials, which travel along the surface of the cell.
- Contractility is the ability of muscle tissue to generate tension when stimulated by an action potential. Tension is the force generated when a muscle contracts that tends to pull the points of attachment at either end toward each other. In an isometric contraction, the muscle develops tension but the length of the muscle remains constant. In an isotonic contraction, the tension developed by the muscle remains almost constant while the muscle shortens.
- Extensibility is the ability of muscle to stretch without being damaged, for example, the heart filling with blood.
- Elasticity is the ability of muscle tissue to return to its original length and shape

after contraction or extension.

A single muscle cell, especially of skeletal muscle, is called a muscle fiber due to its elongated shape. The plasma membrane of the fiber is called the sarcolemma and the cytoplasm is called the sarcoplasm. The sarcoplasm contains an abundance of glycogen and myoglobin, a red protein that binds and stores molecular oxygen. Based on their content of myoglobin, there are two types of skeletal muscle fibers: red muscle fibers, which have a higher myoglobin content, and white muscle fibers, which have a lower myoglobin content.

Skeletal muscles and muscle fibers are surrounded and protected by connective tissue. Each individual muscle fiber is surrounded by a layer of areolar connective tissue called the endomysium. Bundles of ten to one hundred or more muscle fibers, called fascicles, are enclosed by a layer of connective tissue called the perimysium. Fascicles are visible to the naked eye as parallel strands, for example, the "grain" in a cut of meat. The entire muscle is a bundle of fascicles surrounded by yet another layer of connective tissue called the epimysium. The epimysium is continuous with the fascia, which is a sheet or broad band of connective tissue.

There are two types of fascia: deep and superficial. Deep fascia, composed of dense irregular connective tissue, occurs between adjacent muscles. It can separate groups of muscles into muscle compartments, such as occurs in the forearm. Superficial fascia, also known as the subcutaneous layer or hypodermis, is composed of areolar connective tissue and adipose tissue and occurs between the muscles and the skin. When a person "puts on weight" above that desired, it is generally due to the addition of fat to the superficial fascia. The pattern of distribution of this fat is influenced by sex hormones.

Innervation refers to the distribution or supply of nerves to a part of the body. Skeletal muscle tissue is innervated by somatic motor neurons located in the brainstem and spinal cord. Their axons, called somatic motor fibers, divide into about 200 branches at their distal ends. Each branch travels through the endomysium and terminates on the middle portion of a single skeletal muscle fiber, and each muscle fiber is supplied by only one branch from a single neuron. When an action potential arrives at the end of a motor neuron, it stimulates all the muscle fibers supplied by that neuron to contract. Since they all contract at once, one motor neuron plus all the muscle fibers that it innervates is called a motor unit. The average motor unit contains 150 muscle fibers but some have as few as two or three and others have as many as 2,000. The muscle fibers of a single motor unit are not clustered together but are dispersed throughout the muscle and, when stimulated, cause a weak contraction over a wide area.

The area of close proximity between an axon and its target cell is called a synapse. When the second (postsynaptic) cell is a muscle fiber, the synapse is called a neuromuscular junction (NMJ). The axon ends in a swelling called a synaptic knob that is opposite a region of the sarcolemma of the muscle fiber called the motor end plate. A small gap, about 10 nm wide, that separates the two cells is called the synaptic cleft. A neuroglial cell, called a Schwann cell, covers the entire NMJ and isolates it from the surrounding tissue fluid. When an action potential arrives at the NMJ from the first (presynaptic) cell, the axon releases a chemical messenger called a neurotransmitter from secretory organelles called synaptic vesicles. Although many chemicals function as neurotransmitters, the one released at the NMJ is always acetylcholine. Acetylcholine then diffuses across the synaptic cleft and binds to its receptors on the motor end plate. The binding of acetylcholine to its receptors triggers an action potential that travels along the sarcolemma and ultimately causes the muscle fiber to shorten. The muscle fiber relaxes when nervous stimulation ceases.

Myofibrils are the contractile elements of muscle fibers. Myofibrils run the length of a muscle fiber through its sarcoplasm. They consist of numerous parallel myofilaments of two types: thin myofilaments consist of the protein actin and thick myofilaments consist of the protein myosin. The orderly arrangement of actin and myosin myofilaments in myofibrils is responsible for the striations of skeletal muscle fibers. When a muscle fiber is stimulated by a nerve impulse, the actin myofilaments slide past the myosin myofilaments, causing the myofibril, and therefore the muscle fiber, to shorten and thicken. Shortening and thickening of the myofibrils requires the cell to use ATP for energy. This is called the sliding filament model of muscle contraction.

Skeletal muscles are well supplied with nerves and blood vessels. Generally, an artery and one or two veins accompany each nerve that penetrates a skeletal muscle. Blood capillaries supply oxygen and nutrients and remove heat and waste products of muscle metabolism.

Based on structural and functional characteristics, there are three types of skeletal muscle fibers:

- Slow oxidative fibers (also called type I fibers) have the smallest diameter and the highest capacity to generate ATP by aerobic metabolism. They contain large amounts of myoglobin, many mitochondria, have many blood capillaries, and thus are red muscle fibers. They use ATP at a slow rate and, as a result, have a slow contraction velocity. They have a high resistance to fatigue and are adapted for maintaining posture and endurance-type activities, for example, running a marathon.
- Fast oxidative-glycolytic fibers (also called type IIA fibers) have an intermediate diameter and an intermediate capacity to generate ATP by aerobic metabolism. They are relatively rare except in some endurance athletes. They contain large amounts of myoglobin, many mitochondria, have many blood capillaries, and thus are red muscle fibers. They use ATP at a fast rate and, as a result, have a fast contraction velocity. They resist fatigue, but not quite as well as slow fibers, and are adapted for activities such as walking and sprinting.
- Fast glycolytic fibers (also called type IIB fibers) have the largest diameter and the lowest capacity to generate ATP, which occurs by anaerobic metabolism. They contain small amounts of myoglobin, few mitochondria, have few blood capillaries, and thus are white muscle fibers. They use ATP at a fast rate and, as a result, have a fast contraction velocity. They have a low resistance to fatigue and

are adapted for intense movements of short duration, for example, weight lifting or throwing a ball.

Table 6. Types of Skeletal Muscle Fiber

	Red Muscle Fibers	**White Muscle Fibers**
Slow Twitch	**Slow Oxidative** (Type I). Fatigue-resistant; splits ATP slowly; rich blood supply; large stores of myoglobin; many mitochondria for aerobic metabolism. *Examples*: Postural muscles of neck and back.	
Fast Twitch	**Fast Oxidative-Glycolytic** (Type IIA). Splits ATP rapidly; otherwise, same as above. *Examples*: Leg muscles.	**Fast Glycolytic** (Type IIB). Fatigable; splits ATP rapidly; few blood vessels; little myoglobin; large glycogen stores for anaerobic metabolism. *Examples*: Arm muscles.

Most skeletal muscles contain a mixture of all three types of skeletal muscle fibers, but their proportions vary according to the usual action of each muscle. The proportion of different types of muscle fibers in each muscle is mostly determined genetically. The muscle fibers in any one motor unit are all of the same type. The total number of skeletal muscle fibers is generally fixed at birth because skeletal muscle fibers are incapable of mitosis.

Cardiac muscle tissue is found only in the wall of the heart, where it forms the principal tissue. Unlike skeletal muscle tissue, cardiac muscle tissue contracts and relaxes rapidly, continuously, and rhythmically. Also, it can contract without extrinsic stimulation and can remain contracted longer than skeletal muscle tissue. Cardiac muscle cells are joined end to end by intercalated discs, which have three distinctive features:

- Interdigitating folds of the plasma membranes of the adjacent cells, like two layers of corrugated cardboard, increase the surface area of intercellular contact.
- Desmosomes keep the cells tightly joined and prevent them from separating during contraction.
- Gap junctions between the cells act as electrical conductors, allowing action potentials to spread from cell to cell.

Smooth muscle cells can stretch considerably without developing tension. A single cell turns like a corkscrew when it contracts and rotates in the opposite direction when it relaxes. Smooth muscle cells can contract in response to nerve impulses, but they can also contract without nervous stimulation, for example, in response to stretch or hormones. Smooth muscle tissue may be innervated by neuron fibers of either the autonomic nervous system or enteric nervous system. There are two types of smooth muscle tissue:

- Single-unit (visceral) smooth muscle tissue, which is located in the walls of small blood vessels and hollow viscera, contracts as a single unit when one fiber is stimulated. This is due to the presence of gap junctions between cells.
- Multiunit smooth muscle tissue, which is located in the walls of large arteries, bronchioles, arrector pili muscles, and the iris of the eye, is composed of cells that are innervated like those of skeletal muscle. Each motor unit contracts independently of the others upon stimulation.

Growth and regeneration of muscle tissue varies by type. Skeletal muscle grows primarily by hypertrophy (the increase in size of existing cells) because skeletal muscle fibers do not increase in number (hyperplasia) after the first year of life. Since skeletal muscle fibers cannot divide by mitosis, damaged ones are either replaced by those derived from other cell types or replaced by fibrous scar tissue (fibrosis). Like skeletal muscle fibers, cardiac muscle cells can also grow by hypertrophy but cannot divide by mitosis. However, since damaged cardiac muscle cells cannot be replaced by those derived from other cell types nor can they be repaired, they must be replaced by fibrosis. Smooth muscle cells can divide by mitosis, so they can be replaced when damaged. However, they undergo hypertrophy to only a limited extent.

TOPIC 14. THE MUSCULAR SYSTEM

There are two types of attachment between a skeletal muscle and a bone. In a direct attachment, collagen fibers of the epimysium are continuous with the periosteum of the bone – the muscle tissue appears to emerge directly from the bone. An example of this type is the attachment between the masseter muscle and the zygomatic arch. In an indirect attachment, the collagen fibers of the deep fascia are continuous with a tendon that merges into the periosteum of the bone. There are many examples of this type, such as the attachment between the biceps brachii and the scapula. A broad, flat, sheet-like tendon is called an aponeurosis. Examples include the aponeuroses of the internal abdominal oblique and the external abdominal oblique. Instead of to a bone, sometimes the fascia of one muscle attaches to the fascia or tendon of another muscle, or to the collagen fibers of the dermis. Examples of the latter are the skeletal muscles that produce facial expressions.

Skeletal muscles that produce movements do so by pulling on bones or soft structures. Most muscles extend across at least one joint and are attached to the articulating bones that form the joint. When such a muscle contracts, one bone remains relatively stationary while the other (movable) bone is pulled toward it. The attachment of a muscle tendon to the stationary bone is called the origin, or head. The attachment of the other muscle tendon to the movable bone is called the insertion. Many muscles are narrow at the origin and insertion and are thicker in the middle region, which is called the belly of the muscle.

To produce a body movement, bones act as levers and joints function as the fulcrums of these levers. A lever is a rigid rod that moves around some fixed point called a fulcrum (F). The lever is acted on at two different points by two different forces: the resistance (R), which is the force that opposes movement, and the effort (E), which is the force exerted to achieve a movement. Levers are classified by physicists into three types according to which component – fulcrum, resistance, or effort – is in the middle:

- A first-class lever, the most efficient mechanically, has the fulcrum in the middle (EFR), like a seesaw. One of the few first-class levers in the body is the one in which the atlanto-occipital joint is the fulcrum. The resistance is the weight of the face anteriorly. The effort is provided by muscles of the back of the neck that pull down on the nuchal lines of the skull.
- A second-class lever has the resistance in the middle (ERF), like a wheelbarrow. The talocrural (ankle) joint is arguably an example during plantar flexion. In this interpretation, the resistance is the weight of the body on the talus, the fulcrum is the ball of the foot (distal ends of the metatarsal bones), and the effort is provided by the muscles of the lower leg that insert on the calcaneus, elevating the body when they contract. However, some anatomists argue that the talocrural joint is **not** a second-class joint because the calf muscles that pull up on the calcaneus are part of the resistance bearing down on the talus. In their interpretation, the talocrural joint is the fulcrum of a first-class lever, in which the resistance is the Earth and the effort is still provided by the calf muscles, in effect pushing the Earth away from the body when they contract.
- A third-class lever, the least efficient mechanically, has the effort in the middle (FER), like a pair of forceps. Most levers in the human body are third-class, such as the one consisting of the elbow joint, the bones of the forearm, and the biceps brachii muscle.

The movement produced by a muscle is called its action. Most movements of joints require several skeletal muscles acting cooperatively in groups, rather than individually. Muscles can be classified into at least four categories according to their actions, but a given muscle may act differently during different movements of the same joint:

- The prime mover (agonist) is the muscle that is most responsible for the movement during a particular joint action. For example, the prime mover in flexing the elbow is the biceps brachii muscle.
- The antagonist is a muscle that produces an action opposite to that produced by the prime mover. In some cases, it may relax in order to give the prime mover almost complete control over a movement, but usually both muscles function together with the antagonist moderating the speed or range of the prime mover. The prime mover and antagonist together make an antagonistic pair of muscles, such as the biceps and triceps brachii muscles. Such pairs are necessary at joints to produce opposite movements because a given muscle can only pull, not push.

For example, the biceps brachii muscle cannot both flex and extend the elbow. The category to which a muscle belongs depends on the specific movement under consideration. For example, the prime mover in extending the elbow is the triceps brachii muscle.

- The synergist assists the prime mover by reducing unnecessary movement at a joint. For example, the brachialis muscle lies deep to the biceps brachii muscle and assists it as a synergist to flex the elbow.
- The fixator stabilizes the origin of the prime mover so that it can act more efficiently. For example, when the biceps brachii muscle contracts to move the radius, it also pulls on the scapula, which is very loosely attached to the axial skeleton. To prevent lateral movement of the scapula, fixator muscles attached to it contract simultaneously and hold the bone firmly in place.

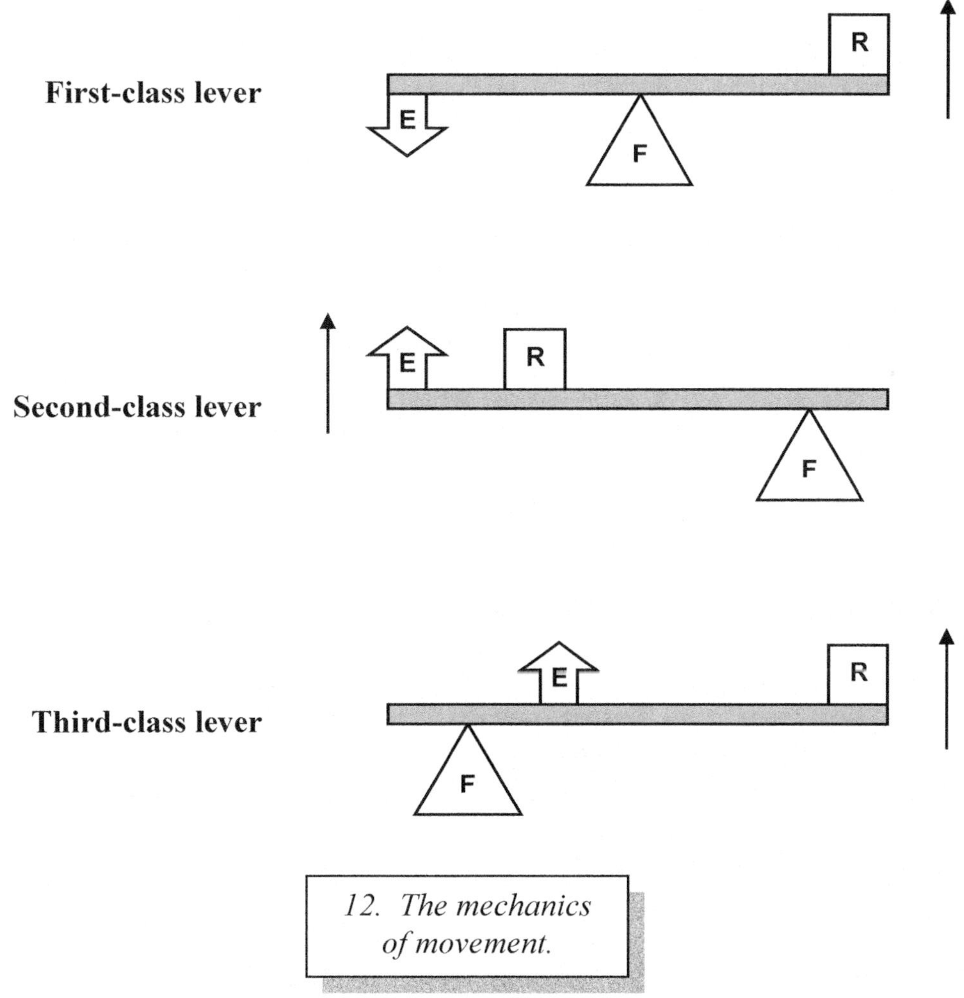

12. The mechanics of movement.

Anatomists distinguish between intrinsic and extrinsic muscles in certain parts of the body, such as the tongue, larynx, back, hand, and foot. An intrinsic muscle has both its origin and insertion within a particular region while an extrinsic muscle has its insertion within a particular region but its origin elsewhere. For example, the intrinsic muscles of the tongue lie entirely within it but the extrinsic muscles of the tongue connect it to the hyoid bone and skull.

Early anatomists named muscles and other structures in their native languages, which quickly led to a state of confusion. Recognizing a need for standard international terminology, in 1895 an international committee of anatomists created a system of Latin names called the *Nomina Anatomica*, which is periodically updated. The customary English names of skeletal muscles are usually only slight modifications of the Latin names. The Latin names of most of the nearly 700 skeletal muscles in the human body are based on a number of anatomical characteristics. These criteria, and examples of muscles that illustrate them, include:

- Direction of muscle fibers. Terms and examples include: **rectus** – fibers parallel to the midline (rectus abdominis); **transversus** –fibers perpendicular to the midline (transversus abdominis); and **oblique** – fibers diagonal to the midline (external oblique).
- Location. Terms and examples include: **frontalis** – of the front (frontalis); **capitis** – of the head (semispinalis capitis); **dorsi** – of the back (latissimus dorsi); **pectoralis** – of the chest (pectoralis major); **abdominis** – of the abdomen (rectus abdominis); **tibialis** – of the tibia (tibialis anterior); **carpi** – of the wrist (flexor carpi radialis); **brachii** – of the arm (triceps brachii); **intercostal** – between the ribs (internal intercostals); **digitorum** – of fingers or toes (flexor digitorum profundus); **pollicis** – of the thumb (adductor pollicis); and **hallucis** – of the big toe (abductor hallucis).
- Size or shape. Terms and examples include: **maximus** – largest (gluteus maximus); **major** – large (zygomaticus major); **medius** – middle-sized (gluteus medius); **minor** – small (zygomaticus minor); **minimus** – smallest (gluteus minimus); **longus** – long (peroneus longus); **brevis** – short (peroneus brevis); **deltoid** – triangular (deltoid); **trapezius** – trapezoidal (trapezius); **serratus** – saw-toothed (serratus anterior); **rhomboideus** – rhomboidal (rhomboideus major); **teres** – round, cylindrical (teres major).
- Relative position. Terms and examples include: **lateral** – farther from the midline (lateral pterygoid); **medial** – closer to the midline (medial pterygoid); **internal** – inside (internal oblique); **external** – outside (external oblique); **superioris** – upper (levator labii superioris); **inferioris** – lower (depressor labii inferioris); **superficialis** – superficial (flexor digitorum superficialis); and **profundus** – deep (flexor digitorum profundus).
- Sites of origin and/or insertion. Terms and examples include: **sternocleidomastoid** – originating on the sternum and clavicle and inserting on the mastoid process of the temporal bone (sternocleidomastoid); and **stylohyoid** – originating on the styloid process of the temporal bone and inserting on the hyoid bone (stylohyoid).
- Number of heads, or origins. Terms and examples include: **biceps** – two heads (biceps femoris); **triceps** – three heads (triceps brachii); and **quadriceps** – four heads (quadriceps femoris).

- Action. Terms and examples include: **flexor** – decreases the angle at a joint (flexor carpi radialis); **extensor** – increases the angle at a joint (extensor carpi ulnaris); **abductor** – moves a bone away from the midline (abductor pollicis brevis); **adductor** – moves a bone closer to the midline (adductor longus); **levator** – raises or elevates (levator scapulae); **depressor** – depresses or lowers (depressor labii inferioris); **supinator** – turns the palm upward or anteriorly (supinator); **pronator** – turns the palm downward or posteriorly (pronator teres); **sphincter** – decreases the diameter of an opening (external anal sphincter); **tensor** – makes tense (tensor fasciae latae); and **rotator** – moves a bone around its longitudinal axis (rotatore, which is a muscle connected to the transverse process of one vertebra and the spinous process of the superior vertebra).

The human muscular system reflects its evolutionary past, as the following vestigial structures demonstrate:

- The auriculares muscles surrounding the external ear (called the auricula or pinna) are vestigial. The auricularis anterior draws the pinna forward and upward, the auricularis superior slightly raises the pinna, and the auricularis posterior draws the pinna backward. In other mammals, these muscles serve to swivel the auricula to point in the direction of interesting sounds, but in humans, all they can manage is a feeble wiggle.
- There is no sensible reason for the existence of the plantaris muscle in the human calf, except a common ancestry with primates. In the primate, it causes all the digits to flex at once, and thus is useful in swinging from trees by the feet. In the human it is atrophied, is absent in 9 percent of the population, and does not even reach the toes but disappears into the calcaneal (Achilles) tendon.

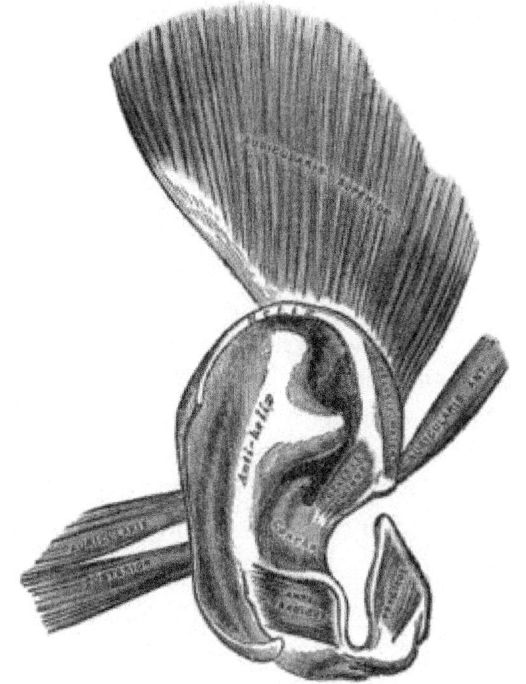

13. The muscles of the auricula.
(Gray's Fig. 906)

- The extensor coccygis muscle of the back is not always present and would wag the tail, if humans had one. This muscle originates on the last segment of the sacrum or the first piece of the coccyx and inserts on the lower part of the coccyx. It is a rudiment of the extensor muscle of the caudal vertebrae of other animals.
- The palmaris longus is a long, narrow muscle that lies superficially in the anterior compartment of the forearm. It originates on the medial portion of the humerus and inserts on the connective tissue of the palm. It helps to flex the wrist and may

once have been important for hanging and climbing from trees. It is missing in 11 percent of modern humans and is so weak and insignificant that surgeons harvest it for reconstructive surgery.[67]

The structure of the abdominal wall is a very good example of an adaptive compromise. The anterolateral abdominal wall consists of skin, layers of fascia, four pairs of muscles, and parietal peritoneum. Each lateral wall contains a set of three flat muscles, consisting (from superficial to deep) of the external oblique, internal oblique, and transversus abdominis. They act to compress the abdominal contents during expiration, micturition, and defecation. The anterior wall contains two rectus abdominis muscles, which extend vertically through the entire length of the wall. Each muscle arises from the pubis and inserts on the lower costal cartilages and xiphoid process (sternum), acting to flex the vertebral column.

The aponeuroses of the external oblique, internal oblique, and transversus abdominis muscles form the rectus sheath, which incompletely encloses the rectus abdominis muscles. The anterior and posterior layers of the sheath meet at the midline, forming the linea albea that connects the two rectus abdominis muscles. In the lower abdominal wall of *Homo sapiens* and many mammalian quadrupeds, the posterior layer of the rectus sheath is absent.

The inguinal ligament is formed by the inferior free border of the external oblique aponeuroses. Just superior and parallel to the medial half of the inguinal ligament is an oblique passageway, about 4 to 5 cm long, through the anterior abdominal wall called the inguinal canal. The inner opening of the inguinal canal is called the deep (abdominal) inguinal ring and the outer opening is called the superficial (subcutaneous) inguinal ring. The canal transmits the spermatic cord in males and the round ligament of the uterus in females. This anatomical arrangement presents no particular functional anatomic difficulty for quadrupeds as their inguinal canal is directed "uphill" during ambulation and therefore is not subjected to significant gravitational stress.[68]

In humans, however, erect posture increases gravitational stress due to the weight of the viscera directed toward the lower abdomen. The canal area is a weak point, subject to protrusions of fat or small intestine, called inguinal hernias, from within the abdominal cavity. In the United States, about 25 percent of males and 2 percent of females have inguinal hernias in their lifetimes, the most common hernia in males and females. If the herniated part becomes strangulated, serious complications, including death, may result. This anatomical arrangement in humans makes no sense if it is attributed to a so-called Intelligent Designer; however, it is easily explained as the result of the evolution of humans from quadrupedal ancestors.

TOPIC 15. WALKING ON TWO LEGS

Several aspects of human anatomy make little sense without recognizing that the human body has a history. Humans and other primates are descendants of certain squirrel-sized, insect-eating (insectivorous) mammals that took up life in the trees about 55 to 60 Mya.

The advantages of living in this arboreal habitat were probably greater safety from predators, less competition, and a rich food supply of insects, lizards, leaves, and fruit. Any new feature that enabled arboreal animals to survive and reproduce more successfully would have been strongly favored by natural selection.

Observing modern monkeys and apes can provide insight into how primates adapt to the arboreal habitat and how certain human adaptations originated. Most primates today inherit these adaptations from their last common ancestor:

- They have a prehensile hand, adapted for grasping branches by encircling them with the thumb and fingers. This facilitates moving from place to place by brachiation (swinging).
- They have stereoscopic vision due to forward-facing eyes, which provides better hand-eye coordination for catching prey and better depth perception for judging distances when leaping from branch to branch.
- Color vision, which is rare among mammals, is the hallmark of primates, enabling them to distinguish ripe, sugary fruits from unripe ones. It also helps them to distinguish between tender young leaves and tough, more toxic older foliage.
- Compared to most other animals, primate brains are large relative to their body size. Those areas of the brain that are involved with controlling manual dexterity, eye-hand coordination, and stereoscopic vision have particularly expanded.

Chimpanzees walk quadrupedally on their knuckles ("knuckle-walk") about 85 percent of the time[69], and in this sense can be considered more terrestrial than arboreal. When they stand on two feet, they do so with sloping backs and bent legs. The chimpanzee has a flat foot, with an opposable hallux and long phalanges for grasping. Nonhuman primates cannot fully extend the knee, nor can they lock the knee when standing. As a result, chimpanzees cannot stand or walk on two legs for very long without tiring. Their femur articulates at the hip and then continues in a straight line downward to the knee joint. Thus, the upper and lower parts of the leg do not meet at an angle. The weight-bearing axis of the leg is medial to the knee joint. As a result, the medial condyle of the femur has a larger surface area than the lateral condyle. In addition, the surface of the tibia is slightly convex, permitting more rotation in the chimpanzee knee joint than in the human.

Chimpanzees lack the extreme curves of the human vertebral column. Because of their **C**-shaped spine, their center of gravity is anterior to the hip joint when they stand; they must exert a continual muscular effort to keep from falling forward, and they fatigue relatively quickly. This is exacerbated by the fact that the forelimbs of apes are longer than their hindlimbs. In fact, sometimes they hold their long forelimbs over their heads when they walk on their hindlimbs. The largest muscle of the buttock, the gluteus maximus, serves in the chimpanzee primarily as an abductor of the thigh – it draws the leg laterally. As a result, the gait of a chimpanzee when it is walking upright is awkward and clumsy. As is typical of quadrupedal mammals, the foramen magnum of the chimpanzee is located toward the posterior of the skull. Due to the weight of the skull anterior to the occipital condyles, the head requires strong muscular attachments to hold it erect. Chimpanzees have prominent supraorbital ridges for the attachment of muscles that pull back on the skull.

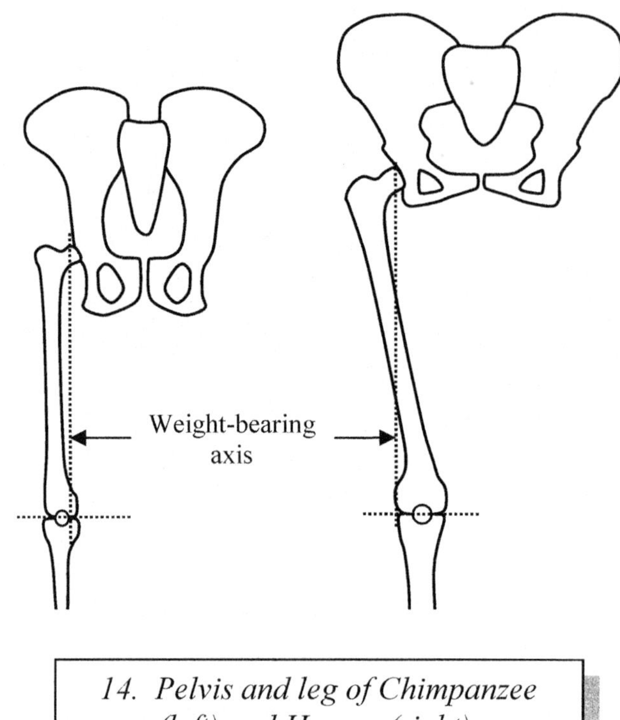

14. *Pelvis and leg of Chimpanzee*
(left) and Human (right).

Only human primates have the ability to walk long distances on two legs. A complete knee joint, consisting of a distal femur and proximal tibia, from a single individual found at Hadar in 1976 was conclusive proof of bipedal walking in the hominin *Australopithecus afarensis*.[70] Footprints preserved in a layer of hardened volcanic ash found at Laetoli in Tanzania in 1978 indicate that the same species walked upright as early as 3.6 Mya.[71]

Habitual bipedal locomotion requires balancing on one foot at a time during walking. The trunk rotates over the supporting limb and foot to help stabilize the upper body as the unsupported leg swings forward. Hip, knee, and ankle joints become fully extended in order to support all of the body weight on one foot. Humans have a number of adaptations for bipedalism:

- Humans have a longitudinal arch in the foot, formed by the tarsals and metatarsals, providing a shock absorbing and weight distribution system for walking.[72] Although their hallux is not opposable as it is in most Old World monkeys and apes, it is highly developed, providing the "toe-off" that pushes the body forward in the last phase of the stride.[73] However, since the human foot is not prehensile, the other digits, especially the fifth (little) toe, have become greatly reduced. In a large number of people the fifth toe has degenerated to a tiny digit often without a nail or only the merest vestige of one.[74] Therefore, the fifth toe is vestigial.

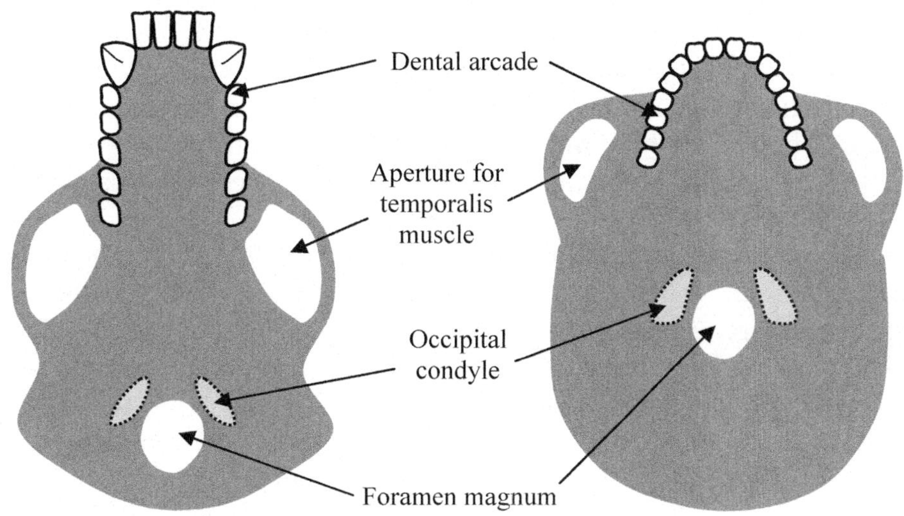

15. Inferior views of cranium of Chimpanzee (left) and Human (right).
(also see <www.eskeletons.org>)

- Humans can not only fully extend the knee, but by a slight medial rotation of the femur on the tibia, they can also lock the knee joint in the fully extended position. In this state, all the major knee ligaments are twisted and taut. This makes it possible to stand erect without tiring the extensor muscles of the leg. To unlock the knee, the popliteus muscle rotates the femur laterally and untwists the ligaments.

- The location of the lower limbs close to the body's midline, under the center of balance, is an adaptation for walking on two legs. As explained by Ian Tattersall, "In bipedal humans…the feet pass close to each other during walking so that the body's center of gravity can move ahead in a straight line. If this didn't happen, the center of gravity would have to swing with each stride in a wide arc around the supporting leg. This would be extremely clumsy and inefficient, wasting a lot of energy. So in bipeds, both femora angle in from the hip joint to converge at the knee; the tibiae then descend straight to the ground. In the human knee joint, this adaptation shows up in the [weight-bearing axis and in the] angle – known as the 'carrying angle' – that is formed between the long axis of the femur and tibia." The human femur angles inward from the hip to the knee joint. The upper and lower parts of the leg meet at an angle. Thus, the weight-bearing axis of the leg is lateral to the knee joint. As a result, the lateral condyle of the femur is larger than the medial condyle. The surface of the tibia is slightly concave for a relatively tight fit at the knee joint.

- The **S**-shaped curve of the vertical human vertebral column is another adaptation for walking on two legs. This is made possible by a greater distance between the pelvis and chest. Robust wedge-shaped intervertebral discs form the lumbar curve. These discs cushion and support the trunk over the pelvis and lower limbs and provide the flexibility of the lumbar region that is critical for trunk rotation.
- Humans have a shortened and broadened ilium, so that the gluteus maximus originates behind the hip joint. Instead of abducting the thigh, the muscle pulls the thigh back in the second half of a stride, which keeps the torso upright and smoothly stable while walking.[75]
- The foramen magnum is located more anterior and the face is much flatter than in the chimpanzee, reducing the weight anterior to the occipital condyles. Thus the head does not require strong muscular attachments to hold it erect. The supraorbital ridges are much lighter and the muscles of the forehead serve only for facial expression, not to hold the head erect.

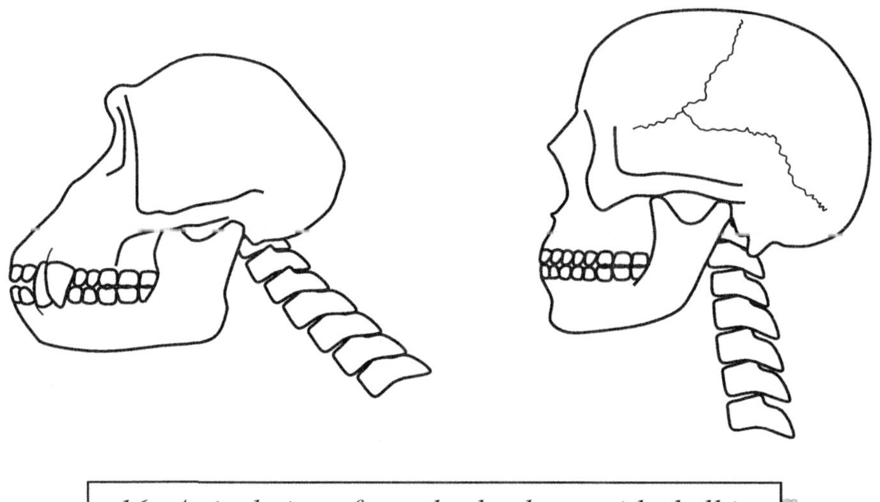

16. Articulation of vertebral column with skull in Chimpanzee (left) and Human (right).

Nevertheless, the anatomy of the knee joint and the lumbar curve of the spinal column in humans are adaptive compromises for bipedal locomotion. Problems occur with the human knee joint because the femur and tibia are not aligned vertically, as they are in the chimpanzee. Furthermore, the lack of a complete, independent articular capsule to connect the bones makes the knee joint rely almost entirely on its associated ligaments and muscles for stability. As a result, the knee joint is the joint in the body most vulnerable to damage. Injuries to the knee joint include dislocation, rupture of the anterior cruciate ligament, damage to synovial membranes, torn menisci, and collateral ligament sprains. In addition, problems develop in the human lower spine because the weight of the top part of the body greatly compresses the lumbar vertebrae, a condition that does not exist in the chimpanzee. Discs in the human lower back can simply wear out (spinal stenosis) with age and cause chronic pain. Intervertebral discs begin to degenerate by the third decade of life. Also, the discs can bulge (herniate) with age and cause chronic pain. Herniated discs are found in a third of adults older than 20, although only 3 percent of these cause symptoms.

TOPIC 16. THE RESPIRATORY SYSTEM

The cardiovascular system and the respiratory system together supply oxygen to and eliminate carbon dioxide from the tissues of the body. Cells need oxygen to obtain energy, releasing carbon dioxide as a byproduct. When carbon dioxide dissolves in the tissue fluid, it produces acidity that is toxic to cells:

$$CO_2 + H_2O \rightleftarrows H_2CO_3 \rightleftarrows H^+ + HCO_3^-$$

The respiratory system enables external (pulmonary) respiration, the process of obtaining oxygen from the environment and eliminating carbon dioxide from the body. The cardiovascular system enables internal (tissue) respiration, the exchange of oxygen and carbon dioxide between the blood capillaries and the tissues. Failure of either the cardiovascular or the respiratory system results in rapid death due to oxygen starvation and accumulation of waste molecules. Other functions of the respiratory system include:

- Housing the olfactory receptors.
- Filtering, humidifying, and warming inspired air.
- Producing sound.
- Helping to eliminate wastes (other than carbon dioxide).

There are two ways of dividing the respiratory system. Structurally, it can be divided into the upper respiratory system, which includes the nose, pharynx, and associated structures, and the lower respiratory system, which includes the larynx, trachea, bronchii, and lungs. Functionally, the respiratory system can also be divided into two parts. The conducting portion consists of the structures that conduct air into the lungs, including the nose, pharynx, larynx, trachea, bronchii, bronchioles, and terminal bronchioles. The respiratory portion consists of tissues within the lungs where gas exchange occurs between air and blood, including the respiratory bronchioles, alveolar ducts, alveolar sacs, and alveoli.

Air enters the respiratory system through the nose. The nose contains the nasal cavity, which is divided into right and left sides by the vertical nasal septum. The anterior portion of the nasal cavity is called the vestibule and opens to the outside through two nares (singular: naris) or nostrils. Posteriorly, the nasal cavity opens to the nasopharynx through two choanae (singular: choana). The nasolacrimal ducts and ducts from the paranasal sinuses open into the nasal cavities. The surface area of the cavity walls is increased by bony projections called conchae (singular: concha); they are covered by a ciliated mucous membrane that filters, warms, and humidifies the air as it passes through the nose. The olfactory receptors are located in the membrane lining the superior nasal conchae and adjacent septum, in a region called the olfactory epithelium. Thus, the major functions of the nose are:

- Warming, moistening, and filtering incoming air.
- Detecting olfactory stimuli.
- Modifying speech vibrations as they pass through the large, hollow resonating chambers.

The pharynx (throat) is a funnel-shaped tube that is located anterior to the cervical vertebrae and posterior to the nasal cavity, oral (buccal) cavity, and larynx. Its wall is composed of skeletal muscles and lined with mucous membrane. The pharynx can be divided into the nasopharynx, the oropharynx, and the laryngopharynx. The nasopharynx is the uppermost portion that extends from the choanae to the plane of the soft palate. The oropharynx is the middle portion that extends from the plane of the soft palate to the level of the hyoid bone and opens to the oral cavity through the fauces. The laryngopharynx or hypopharynx is the lowermost portion that extends from the level of the hyoid bone downward to the esophagus posteriorly and the larynx anteriorly. It opens to the larynx through the glottis. The major functions of the pharynx are:

- Conducting air and food.
- Providing a resonating chamber for speech sounds.
- Housing the tonsils, which participate in immunological defense.

The larynx (plural: larynges), or voice box, is a short passageway that extends from the pharynx to the trachea. Most of the larynx is lined by a mucous membrane containing ciliated pseudostratified columnar epithelium that moves dust-laden mucus upward toward the pharynx. The wall of the larynx contains nine pieces of cartilage, including the thyroid cartilage and the epiglottis. The thyroid cartilage, or Adam's apple, is composed of two fused plates of hyaline cartilage that form the anterior wall of the larynx. The epiglottis is a large leaf-shaped piece of elastic cartilage, the stem of which is attached to the anterior rim of the thyroid cartilage. During swallowing, its free edge closes the glottis and prevents food and liquids from entering the larynx. The vocal folds, or true vocal cords, are two folds of mucous membrane supported by elastic ligaments stretched across the glottis. When air passes through the glottis, these folds can vibrate, producing sound. This process is known as phonation.

Humans are unusual (but not unique[76]) among mammals in having a permanently "descended larynx." The resting location of the standard mammalian larynx is high in the throat and typically situated in the nasopharynx, allowing these animals to swallow fluids and breathe through the nose simultaneously, but it does not allow them to breathe through the mouth. For example, a dog has to make a special effort to bring its larynx down into its throat in order to bark or to pant; when it relaxes, the larynx goes back up again. In the chimpanzee, the tongue is positioned entirely within the oral cavity, and the larynx is positioned high, close to the choanae (see Figure 17). The epiglottis and soft palate overlap to form a watertight seal when the larynx is raised, locking into the nose during feeding. In other words, the larynx pokes up into the nasal passage like a snorkel so chimps can continue to breathe while they eat. The hyoid bone is connected to the larynx, mandible, and skull by means of muscles and ligaments; it is part of the anatomic system that can raise the larynx. This elevated position also typifies human neonates, making it possible for them to breathe and nurse at the same time.

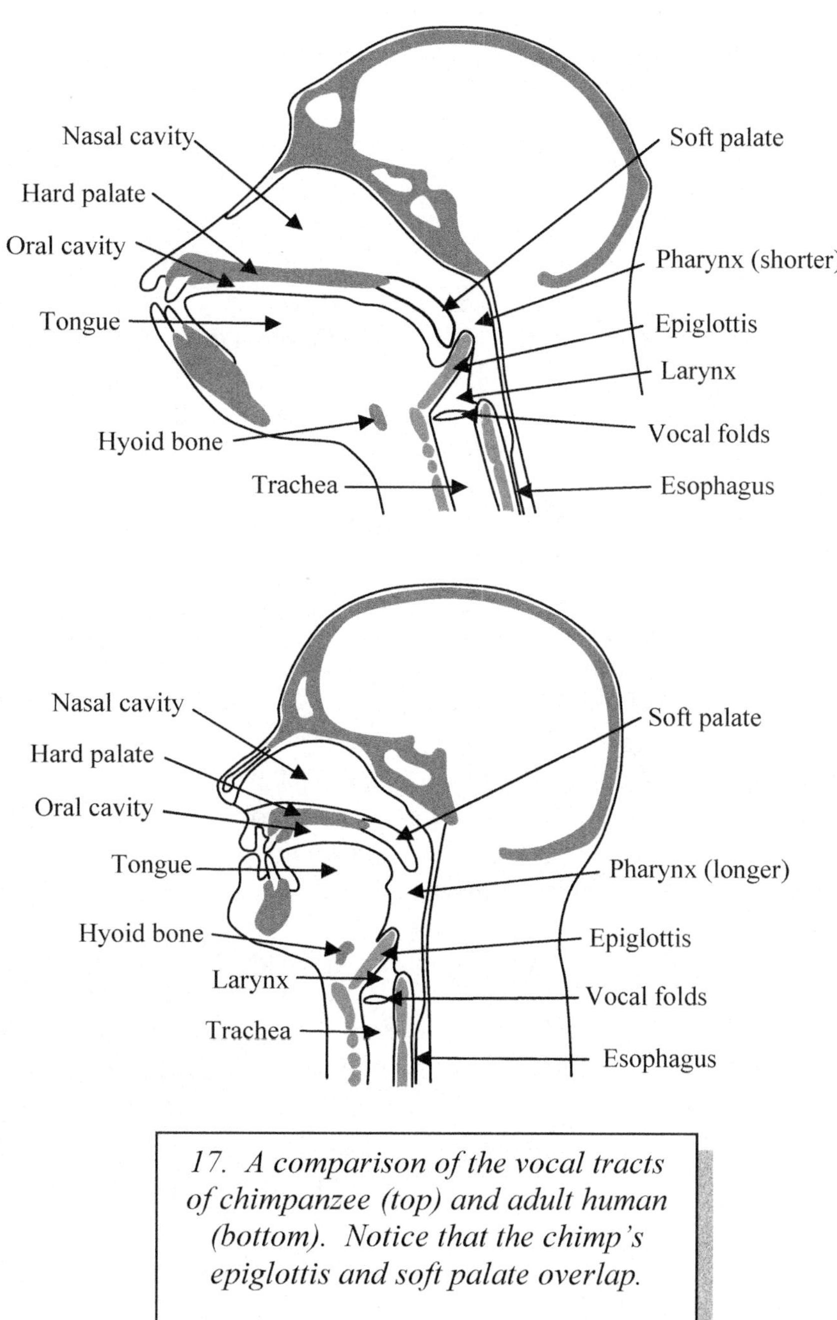

17. A comparison of the vocal tracts of chimpanzee (top) and adult human (bottom). Notice that the chimp's epiglottis and soft palate overlap.

The human larynx typically begins to gradually move lower at about three months of age and reaches its final position by age four. This makes articulate speech possible because air passing through the oral cavity can thus be modified by the tongue and lips. Modeling and simulation studies have shown, however, that the limited phonatory capabilities of the high-positioned larynx in primates and babies represent only a relatively minor handicap in terms of language.[77] For that matter, the high position of the larynx in human babies does not prevent them from imitating the adult vowel sounds "ee", "ah", and "oo" from as early as four months of age, and from producing their first words eight months later, when the larynx is still very high and the pharyngeal cavity is still very small. Therefore, the reason that apes and younger babies cannot speak is apparently not that their larynx is too high, but rather that they lack the cognitive abilities needed to master language.

Thus, it seems unlikely that the descended larynx in humans was "selected for" language. It may have conferred some evolutionary advantages in pre-linguistic communication, and/or it may have afforded certain benefits with regard to breathing. Since other animal species besides humans (deer, for example) also have low larynges, this anatomical characteristic may have evolved because it lets animals make sounds that lead others to believe that they are larger than they really are. By 350,000 years ago, human ancestors had evolved a descended larynx. But some experts say that spoken language probably developed much later – less than 100,000 years ago. This would make the human larynx an example of an exaptation – in other words, an adaptation to pressures selecting for purposes other than speech, but whose result (that is, a descended larynx) nevertheless facilitated the articulation of words.[78]

In any case, as a result of the descended larynx, the respiratory and digestive tracts now cross each other in the area of the pharynx because the epiglottis and soft palate can no longer overlap. The major problem of this new configuration is that a bolus of food can become lodged in the entrance of the larynx, and if this material cannot be expelled rapidly, an individual may literally choke to death. As Charles Darwin wrote, "...every particle of food and drink which we swallow has to pass over the orifice of the trachea, with some risk of falling into the lungs, notwithstanding the beautiful contrivance by which the glottis is closed."[79] Another disadvantage of the crossed pathways is the relative ease with which vomit can be aspirated into the trachea, and thus pass into the lungs. Hence, the descended human larynx is an example of an adaptive compromise (a very poor design).

The trachea, or windpipe, is a tube located anterior to the esophagus that extends from the larynx down to the level of the fifth thoracic vertebra where it divides into right and left primary bronchi. It is lined by a mucous membrane containing ciliated pseudostratified columnar epithelium that moves dust-laden mucus upward toward the pharynx. The wall contains a stack of 16 to 20 **C**-shaped pieces of hyaline cartilage whose open posterior sides face the esophagus. These incomplete rings of cartilage permit slight expansion of the esophagus into the trachea during swallowing. They also provide a rigid support to prevent inward collapse of the tracheal wall.

The lungs are large, cone-shaped organs located in the thoracic cavity. They extend from the thoracic diaphragm (the base) to just slightly above the clavicles (the apex) and lie against the ribs anteriorly and posteriorly. The left lung, which is somewhat smaller than the right, has a superior lobe and an inferior lobe. The cardiac notch, against which the heart rests, is located on its medial surface. The right lung has a superior lobe, a middle lobe, and an inferior lobe. The two lungs are separated by the heart and other structures in the mediastinum. Two continuous layers of a serous membrane, called the pleural membrane, or pleura, enclose and protect each lung. The outer layer, called the parietal pleura, lines the thoracic wall and diaphragm. The inner layer, called the visceral pleura, covers the lungs themselves. The pleural cavity is the narrow gap between the visceral and parietal pleurae and contains a lubricating pleural (serous) fluid secreted by the membranes.

The continuous branching of airways in each lung forms a bronchial tree. The tree begins with a primary bronchus, which is constructed similarly to the trachea. The right primary bronchus is shorter and wider than the left. Each primary bronchus enters a lung at its hilus, a recessed region through which nerves, ducts, and blood vessels pass into an organ, and then divides into secondary (lobar) bronchi. Since each secondary bronchus serves a different lobe of the lung, there are three in the right lung and two in the left. The secondary bronchi divide into tertiary (segmental) bronchi, each of which serves a different segment of a lobe. Blood is supplied to the bronchi of each lung by a bronchial artery. The vein that drains the bronchi of each lung is a bronchial vein. Both blood vessels are part of the systemic circulation. Finally, the tertiary bronchi divide into bronchioles, each of which branches into 50 to 80 terminal bronchioles.

Several structural changes occur as the branching becomes more extensive in the bronchial tree. The epithelium changes from ciliated pseudostratified columnar in the bronchi to ciliated simple columnar in larger bronchioles to mostly ciliated simple cuboidal in smaller bronchioles to mostly nonciliated simple cuboidal in terminal bronchioles. The **C**-shaped rings of cartilage in the primary bronchi are gradually replaced by plates of cartilage that finally disappear in the distal bronchioles. As the amount of cartilage decreases, the amount of smooth muscle tissue increases. The bronchiole wall contains bundles of smooth muscle, whose contraction can close off these airways. This is what happens during an asthma attack, which can be a life-threatening event.

Each terminal bronchiole divides into two or more microscopic respiratory bronchioles. Each respiratory bronchiole divides into 2 to 10 elongated, thin-walled airways called alveolar ducts that end in irregularly shaped spaces called alveolar sacs. Alveoli bud from the walls of the respiratory bronchioles, alveolar ducts, and alveolar sacs. The presence of alveoli defines the respiratory portion of the respiratory system because gas exchange occurs primarily at the alveoli.

A pulmonary alveolus is a cup-shaped evagination lined by simple squamous epithelium and supported by a thin elastic basement membrane. Each alveolus is surrounded by a network of capillaries that receive blood from pulmonary arterioles and that are part of the pulmonary circulation. The alveolar wall and the capillary wall together form the respiratory (alveolar-capillary) membrane through which O_2 and CO_2 diffuse. Certain cells in the alveolar wall secrete surfactant, which lowers surface tension and prevents

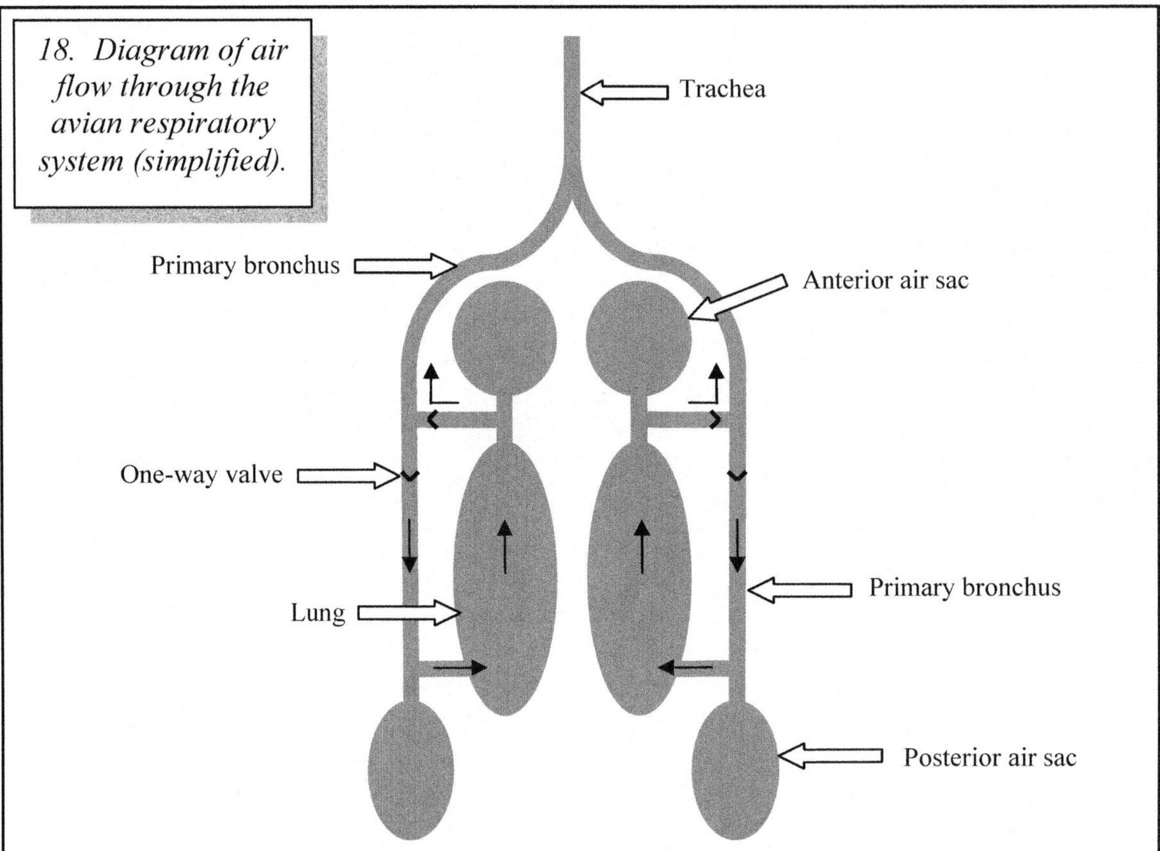

18. Diagram of air flow through the avian respiratory system (simplified).

Like amphibians and all mammals, humans have a tidal pattern of pulmonary ventilation. Air flows in and out of the lungs through a single, two-way path, similar to the ebb and flow of the tide. Because the alveoli cannot completely collapse at the end of expiration, a quantity (called the residual volume) of unexpelled, stale dead air filled with carbon dioxide remains in the lungs at all times. This dead air mixes with the fresh air that enters from the outside, so that the air in the alveoli is never fully oxygenated. Thus, tidal pulmonary ventilation is an inefficient design.

In contrast, air flows through the lungs of birds in one direction only, from posterior to anterior. The avian respiratory system consists of the trachea, bronchi, numerous air sacs, and lungs. The air sacs are **not** involved in gas exchange, and the lungs have a **fixed** volume. During inhalation, all air sacs expand as inhaled air enters the posterior air sacs and lungs and, simultaneously, air moves out of the lungs and into the anterior air sacs. During exhalation, the air sacs diminish in volume as air moves (1) from the posterior air sacs through the lungs and (2) from the anterior air sacs and out of the body via the trachea.[80] There is no residual volume, so the air passing across the respiratory epithelium of the avian lung is always fully oxygenated. Thus, birds can extract about 5 percent more oxygen from a given volume of inhaled air than humans can.

A great deal of evidence has accumulated over the last 30 years that supports the hypothesis that birds evolved from dinosaurs. A recent analysis shows the presence of a very bird-like pulmonary system in predatory dinosaurs.[81] In these dinosaurs, a portion of the air sac actually integrates with the skeleton, forming air pockets in otherwise dense bone. The exact function of this skeletal modification is not completely understood, but the skeletal air pockets may have evolved in these dinosaurs because it lightened the bone structure, allowing them to walk upright. Or, it may have evolved because it allowed them to extract sufficient oxygen from an atmosphere with reduced oxygen levels.[82] Another possible reason might have been that it enabled them to maintain a stable and high metabolism, putting this group of dinosaurs much closer to a warm-blooded existence. In any case, birds have a more efficient respiratory system than humans do because the human body has a different evolutionary history.

alveoli from collapsing during expiration. Associated with the alveolar wall are alveolar macrophages, or dust cells, which are wandering phagocytes that remove dust and other debris.

Pulmonary ventilation (breathing) is the inspiration and expiration of air between the atmosphere and the lungs. Inspiration (inhalation) is the process of moving air into the lungs, whereas expiration (exhalation) is the process of moving air out of the lungs.

The major inspiratory muscles are the thoracic diaphragm (the most important) and the external intercostals. The process of inspiration involves several steps. First, contraction of the inspiratory muscles increases the volume of the thoracic cavity. Increased volume causes a decrease in the pressure within the pleural cavity (intrapleural pressure). Decreased intrapleural pressure "pulls" the walls of the lungs outward. Increased volume of the lungs causes a decrease in the pressure within the alveoli (alveolar pressure). Therefore, air rushes down the pressure gradient from the atmosphere into the lungs. Inspiration ends when the alveolar pressure becomes equal to the atmospheric pressure. Forceful inspiration involves the contractions of accessory muscles of inspiration, including the sternocleidomastoids, scalenes, and pectoralis minors.

Unlike inspiration, normal expiration is a passive process since no muscular contractions are necessary. The process of expiration also involves several steps. First, the diaphragm relaxes and the abdominal organs press up against it, while the thoracic cage moves down and inward as the external intercostal muscles relax. The above movements decrease the volume of the thoracic cavity. Decreased volume causes an increase in the intrapleural pressure. Increased intrapleural pressure pushes the walls of the lungs inward. The volume of the lungs decreases as well as that of the alveoli within them, and the alveolar pressure rises. Therefore, air rushes down the pressure gradient from the lungs into the atmosphere. Forceful expiration involves the contractions of abdominal and internal intercostal muscles.

Topic 17. An Overview of the Circulatory System

The circulatory system has two components: the cardiovascular system and the lymphatic/immune system. The cardiovascular system consists of the heart, blood vessels, and blood. Arteries conduct blood from heart to tissues, and veins return blood from tissues to heart. Small arteries called arterioles are joined to small veins called venules by blood capillaries, which form networks within the tissues. Physiological exchange between blood and tissues takes place through the thin walls of blood capillaries.

Water and small solutes leak out of blood capillaries driven by dialysis, osmosis, and the fluid (hydrostatic) pressure produced by the heart and in some places by gravity. Fluids would accumulate in the tissues and cause swelling (edema) were they not drained away by the second component of the circulatory system, the lymphatic system. It consists of lymphatic vessels and lymphatic capillaries, the latter of which end blindly in the tissues. Tissue fluids enter the lymphatic capillaries, where they constitute lymph. Lymph passes slowly into larger and larger lymphatic vessels until it is eventually discharged into the cardiovascular system.

TOPIC 17.1. THE CARDIOVASCULAR SYSTEM: THE BLOOD

Blood tissue is a connective tissue with a fluid extracellular matrix. It has several major physical characteristics. Blood is more viscous than water and is slightly adhesive. Its temperature in the body is 38° C and its pH ranges between 7.35 and 7.45. It constitutes 8 percent of total body weight, with a volume of 5 to 6 liters in males and 4 to 5 liters in females. Blood performs three major functions:

- Transportation of oxygen, carbon dioxide, nutrients, heat, wastes, and hormones.
- Regulation of pH, body temperature, and water content of cells.
- Protection against blood loss via clotting, and against foreign microbes and toxins via the action of phagocytic white blood cells and specialized plasma proteins.

Blood is composed of plasma and formed elements. Blood plasma, or plasma, comprises about 55 percent of the volume of blood. It is a straw-colored liquid that is composed of water (91.5 percent) and solutes (8.5 percent). The solutes include plasma proteins (albumins, globulins, and fibrinogen), waste molecules, nutrient molecules, enzymes, hormones, gases, and electrolytes. Formed elements comprise the remaining 45 percent of the volume of blood. They do not dissolve in the plasma and consist of cells and cell derivatives. Greater than 99 percent of the formed elements in blood are erythrocytes, also called red blood cells, or RBCs. In addition, leukocytes, also called white blood cells or WBCs, and platelets are formed elements. The production of blood cells of all kinds is called hemopoiesis, or hematopoiesis, and begins after birth in the myeloid tissue of red bone marrow.

Each mature erythrocyte is a flexible, biconcave disc that lacks a nucleus or any other organelle. Due to the lack of organelles, it cannot replace plasma membrane components that are damaged as it squeezes through narrow capillaries, so each erythrocyte lasts only about 120 days. A healthy male has about 5.4 million erythrocytes per µl of blood and a healthy female 4.8 million. Erythropoiesis is the formation of erythrocytes that occurs in red bone marrow. Hypoxia, which is a lack of oxygen in body tissues, stimulates the kidneys to release erythropoietin, which in turn stimulates erythropoiesis. Each erythrocyte contains about 200 million hemoglobin molecules dissolved in its cytosol. Each hemoglobin molecule consists of four globin polypeptide chains and four heme groups containing iron. Hemoglobin can bind reversibly to oxygen (O_2), permitting erythrocytes to transport oxygen from the lungs to other tissues of the body. Hemoglobin can also bind reversibly to carbon dioxide, a waste molecule, permitting erythrocytes to transport carbon dioxide from the tissues to the lungs to be expelled. Hemoglobin is also involved in the regulation of blood pressure.

Erythrocytes lack the surface proteins called major histocompatability (MHC) antigens that nucleated body cells have. MHC antigens are cell-identity markers that identify the cells of the body to its immune system. The immune system destroys cells with "nonself" MHC antigen but ignores cells with "self" antigen or no antigen. This is why blood transfusions between individuals are possible.[xiii] There are two major blood group systems, the ABO blood group system and the Rh blood group system. Each is based on the presence or absence of genetically determined RBC-surface antigens called agglutinogens. These are the antigens that can cause blood type incompatibilities in blood transfusions or between mother and fetus.

Leukocytes have a nucleus but do not contain hemoglobin. They are usually larger than erythrocytes. Unlike erythrocytes, which are confined to the blood, leukocytes are also found in lymph and tissue fluid. The lifespans of the various types of leukocytes range from several hours to many years, but most WBCs live for only several hours to several days. The average number of leukocytes ranges from 5,000 to 10,000 per μl of blood. Therefore, the ratio of RBCs to WBCs is about 700 to 1. The two major classes of WBCs are the granular leukocytes and the agranular leukocytes.

Granular leukocytes or granulocytes have lobed nuclei and prominent granules in their cytosol. There are three types: neutrophils, eosinophils, and basophils. Neutrophils have very fine granules, some of which stain with acidic dyes and others with basic dyes. The combined effect gives their cytoplasm a pale lilac color. They are phagocytes that normally constitute 60 to 70 percent of the leukocyte count. Neutrophils are also known as polymorphonuclear leukocytes due to the variation in the shapes of their nuclei. Eosinophils have coarse granules that stain with the acidic dye called eosin, giving their cytoplasm a pinkish orange to rosy color. These cells release antihistamines in allergic reactions and are also effective against certain parasitic worms. They normally make up 2 to 4 percent of the leukocyte count. Basophils have coarse granules that stain with basic dyes, giving their cytoplasm a dark purple color. These cells release substances, such as histamine (a vasodilator) and heparin (an anticoagulant), which intensify the inflammatory reaction. Their frequency ranges from less than ½ percent to 1 percent of the leukocyte count.

Agranular leukocytes or agranulocytes have poorly visible granules in their cytosol. There are several types, including monocytes and lymphocytes. Monocytes are the largest leukocytes, the nuclei of which may be ovoid, kidney-shaped, or horseshoe-shaped. They make up 3 to 8 percent of the leukocyte count. In the tissues, they differentiate into even larger macrophages. Lymphocytes are about the same size as erythrocytes but have a prominent rounded nucleus, which may nearly fill the entire cell. Their frequency ranges from 25 to 33 percent of the leukocyte count. There are several subclasses of lymphocytes with different functions, but they all look alike through the light microscope. These subclasses include: B-lymphocytes, or B-cells, which differentiate into plasma cells that produce antibodies; T-lymphocytes, or T-cells, which destroy foreign cells and help other cells of the immune system; and natural killer (NK) lymphocytes, which attack body cells that are infected with viruses or that are cancerous.

[xiii] An antigen is a substance that stimulates the production or mobilization of antibodies. An antibody is a protein manufactured by the body that binds to an antigen on a bacterium, toxin, or extracellular virus. This flags them for subsequent destruction by the immune system.

Like erythrocytes, platelets do not have nuclei. They are membrane-enclosed structures, about half the size of erythrocytes, formed from the fragmentation of cells called megakaryocytes. Nevertheless, they contain lysosomes, endoplasmic reticulum, a Golgi complex, and vesicles; they are also capable of amoeboid movement and phagocytosis. Platelets help repair slightly damaged blood vessels by promoting blood clotting. The average number of platelets in the blood ranges from 130,000 to 360,000 per μl. Each platelet lasts from five to nine days.

TOPIC 17.2. THE CARDIOVASCULAR SYSTEM: THE HEART

The central component of the cardiovascular system is the heart, the function of which is to propel blood through the blood vessels. The blood vessels form two closed circuits called the systemic circulation and the pulmonary circulation. The systemic circulation enables internal respiration; it carries blood from the heart to all capillaries of the body except the pulmonary capillaries and then back to the heart. The pulmonary circulation enables external respiration; it carries blood from the heart to the pulmonary capillaries and then back to the heart. The heart is really two pumps in one. The left side of the heart pumps oxygenated blood into the systemic circulation and the right side of the heart pumps deoxygenated blood into the pulmonary circulation.

The heart is a hollow, cone-shaped organ that is about the size of a person's closed fist – about 12 cm long, 9 cm wide, and 6 cm thick. Its lower, pointed end is called the apex and its broader, upper portion is called the base. It is located between the lungs in the mediastinum, and about two-thirds of the heart is left of the body's midline.

The pericardium surrounds and protects the heart. It consists of two major layers: the outer fibrous pericardium and the inner serous pericardium. The outer fibrous pericardium is composed of a tough, inelastic, dense irregular connective tissue. It prevents overstretching of the heart, provides protection, and anchors the heart in the mediastinum. The inner serous pericardium is a serous membrane that consists of two continuous layers. The outer parietal layer is fused to the fibrous pericardium and the inner visceral layer, or epicardium, adheres tightly to the surface of the heart. The narrow gap between the two layers, called the pericardial cavity, contains 5 to 30 ml of pericardial (serous) fluid that reduces friction between the membranes as the heart moves.

The wall of the heart consists of three layers: the outer epicardium, the middle myocardium, and the inner endocardium. As mentioned above, the epicardium is essentially the visceral serous pericardium, which over much of the heart is associated with thick deposits of adipose tissue that fill grooves in the heart surface and protect the coronary blood vessels. The thickest layer by far is the myocardium, which is composed of cardiac muscle tissue that does the work of the heart. It contains a meshwork of collagenous and elastic fibers that form the fibrous skeleton of the heart. The fibrous skeleton provides structural support for the heart and a base for the attachment of its valves, gives its muscle something to pull against, and acts as an electrical insulator for its conduction system. Since the heart has a high metabolic rate, requiring an abundant supply of oxygen and nutrients, the myocardium has an extensive network of blood vessels, called the coronary circulation, to ensure that fresh oxygenated blood quickly reaches every muscle cell. The endocardium is composed of a layer of thin areolar connective tissue covered by endothelium (simple squamous epithelium). The endothelium is continuous with the endothelium of the blood vessels.

The heart has four chambers. The two superior chambers are the right atrium and left atrium, which receive blood returning to the heart. Each has an appendage called an auricle that slightly increases its volume. The atria are separated by the interatrial septum, which has an oval depression called the fossa ovalis. The two inferior chambers are the right ventricle and left ventricle, which are separated by the interventricular septum. The right and left ventricles pump blood into the pulmonary trunk and the aorta, respectively. The thickness of the myocardium of the four chambers varies according to function. The walls of the atria are thinner than the walls of the ventricles, since the atria pump blood only to the ventricles immediately below. The walls of the left ventricle, which pumps blood through the entire body, are thicker than the walls of the right ventricle, which pumps blood only to the lungs and back.

Four valves in the heart ensure that blood flows in only one direction. Each consists of two or three cusps – flaps that are composed of dense connective tissue covered by endocardium. The tricuspid valve, located between the right atrium and the right ventricle, and the bicuspid (mitral) valve, located between the left atrium and the left ventricle, are together known as the atrioventricular (AV) valves. The pulmonary valve, located between the right ventricle and the pulmonary trunk, and the aortic valve, located between the left ventricle and the aorta, are together known as the semilunar (SL) valves. String-like chordae tendineae, like the shroud lines of a parachute, connect the cusps of the AV valves to conical papillary muscles on the floor of the ventricle. When the ventricles contract, they force blood against the valves, pushing the cusps together, which seals the openings and prevents blood from flowing back into the atria. At the same time the myocardium contracts, so do the papillary muscles, pulling the chordae tendineae taut and preventing the AV valves from prolapsing (opening backwards) into the atria.

Blood in the right side and left side of the heart is kept entirely separate. Three veins deliver deoxygenated blood to the right atrium. The superior vena cava drains blood from most parts of the body superior to the heart, the inferior vena cava from all parts of the body inferior to the thoracic diaphragm, and the coronary sinus from the wall of the heart. The right ventricle receives blood from the right atrium and pumps it into a large artery called the pulmonary trunk, which immediately divides into the right and left pulmonary arteries. These transport blood to the alveoli, where carbon dioxide leaves and oxygen enters the blood. Two pulmonary veins from each lung (a total of four veins) deliver freshly oxygenated blood to the left atrium. The left ventricle receives blood from the left atrium and pumps it into the ascending aorta, which gives rise to the right and left coronary arteries that deliver blood to the wall of the heart.

The gap junctions of the intercalated discs between cardiac muscle cells permit the rapid transmission of action potentials from one cell directly to the next. However, the cells form two separate functional networks, one within the muscular walls and partition of the atria and the other within the muscular walls and partition of the ventricles. Each network normally contracts as a functional unit in an orderly sequence: first the atria contract, and then the ventricles contract.

The orderly sequence in which the myocardium contracts is maintained by a conduction system consisting of cardiac muscle cells that are specialized to produce and conduct action potentials. Although all cardiac muscle cells are autorhythmic, action potentials delivered to contractile cells by cells of the conduction system stimulate them to contract in the proper sequence. Also, the conduction system acts as a pacemaker to set the rate of contraction of the entire heart. However, various hormones and neurotransmitters can speed or slow pacing of the heart. The conduction system consists of the sinoatrial (SA) node, the atrioventricular (AV) node, the atrioventricular (AV) bundle, the bundle branches, and the conduction myofibers.

The SA node is located in the wall of the right atrium near where the superior vena cava enters it. (The name sinoatrial is derived from the name of this area in the embryo: sinus venosus.) Because the action potentials from the SA node are produced at a greater frequency than in any other part of the conduction system, the SA node sets the frequency of the heartbeat. As the action potential it produces spreads through the muscle cells of the atria, it causes them to contract from the area of the SA node toward the ventricles. However, the wave of excitation is prevented from spreading to the ventricles by the fibrous skeleton, acting as an electrical insulator, in the partition between the atria and ventricles.

The AV node is located on the right side of the interatrial septum just superior to the ventricles. An action potential from the SA node causes the AV node to produce an action potential that leaves the latter along a route, called the AV bundle or bundle of His, in the posterior part of the interatrial septum. The AV bundle divides into the right and left bundle branches, which enter the interventricular septum and descend through it toward the apex of the heart. As each bundle branch approaches the apex, it forms conduction myofibers or Purkinje fibers. At the apex, the conduction myofibers turn upward and branch throughout the ventricular myocardium.

The cardiac cycle comprises all the events associated with one heartbeat. Systole is the contraction phase of a heart chamber and diastole is the relaxation phase. In a normal cardiac cycle, both atria contract while both ventricles relax, and both ventricles contract while both atria relax. The cardiac cycle can be divided into three major phases: the relaxation period, atrial systole, and ventricular systole. During the relaxation period (0.40 sec), all four chambers of the heart are in diastole. The ventricles expand as they relax, passively drawing blood from the atria through the atrioventricular valves. About 75 percent of ventricular filling occurs in this phase, just after the AV valves open. During atrial systole (0.15 sec), the remainder of ventricular filling (25 percent) occurs. During ventricular systole (0.30 sec), both SL valves open and the AV valves close.

TOPIC 17.3. THE CARDIOVASCULAR SYSTEM: THE BLOOD VESSELS

Arteries conduct blood away from the heart and are constructed to withstand the surges in blood pressure produced during ventricular systole. The walls of arteries consist of three layers, or tunics. The innermost layer, called the tunica interna (intima), surrounds the lumen. It consists of a simples squamous endothelium overlying a basement membrane and a sparse layer of fibrous connective tissue, called the internal elastic lamina. The middle layer, called the tunica media, is usually the thickest layer. It consists of smooth muscle, elastic tissue, and collagen. Contraction of the smooth muscle causes vasoconstriction and relaxation causes vasodilation. The outermost layer is called the tunica externa, or tunica adventitia. It consists of loose connective tissue that often is continuous with the adventitia of adjacent blood vessels, nerves, or organs. Sometimes, an external elastic lamina lies between the tunica externa and tunica media.

Arteries are classified by diameter into three types, although of course there is continuous variation in the diameters of arteries:

- The largest are conducting, or elastic, arteries such as the pulmonary arteries, aorta, and common carotid arteries. The elastic tissue in their tunica media allows conducting arteries to expand when the ventricles eject blood during systole and to recoil during diastole, thereby reducing fluctuations in blood pressure farther along the bloodstream. About one-third of the wall of an elastic artery consists of smooth muscle.
- Smaller branches farther away from the heart that distribute blood to specific organs are called distributing arteries, such as the brachial, femoral, and radial arteries. Because smooth muscle makes up about three-fourths of the wall of a distributing artery, in the form of 25 to 40 layers of muscle cells, these are also known as muscular arteries.
- Resistance, or small, arteries have up to 25 layers of smooth muscle cells and relatively little elastic tissue. Compared to larger arteries, their tunica media is thicker in proportion to the lumen. These arteries are usually too numerous and variable in location to be given names.

An arteriole is a very small branch of an artery with only one or two layers of smooth muscle cells in its tunica media. Metarterioles are short vessels that link arterioles and capillaries. Unlike arterioles, metarterioles have a discontinuous tunica media, that is, they have individual muscle cells spaced a short distance apart.

The microcirculation is that part of the circulatory system concerned with the exchange of gases, fluids, nutrients, and metabolic waste products. Exchange between the blood and tissue fluid by diffusion occurs mainly at the capillaries, and their structure reflects their function. Consequently, capillary walls consist of only two thin layers: an endothelium and a basement membrane.

Capillaries can be classified according to the sizes of gaps between or through the endothelial cells in their walls. The most common type of capillary has a continuous endothelium, in which the endothelial cells form an uninterrupted lining held together by tight junctions. Intercellular clefts about 4 nm wide exist between the endothelial cells that permit small solutes, such as glucose, but not formed elements, plasma proteins, and other large molecules to pass into the tissues. The continuous capillaries of the brain lack intercellular clefts and have more complete tight junctions, forming the blood-brain barrier. Some capillaries have a fenestrated endothelium, in which the endothelial cells contain numerous large holes called fenestrations, or filtration pores, each covered by a thin mucoprotein diaphragm. The fenestrations permit especially rapid movement of small molecules into the tissues but still retain proteins and larger particles in the blood. The fenestrated capillary is important in parts of the body that engage in rapid absorption or filtration, for example, the small intestine and kidneys.

A capillary network, or bed, consists of a group of 10 to 100 capillaries that are supplied by a single metarteriole. There is not enough blood in the body to fill the entire cardiovascular system at once; consequently, about 75 percent of the body's capillaries are closed at any given time. The contraction and relaxation of smooth muscle fibers in the wall of a metarteriole help regulate but do not completely stop blood flow into the capillary network. However, there is a simple ring of smooth muscle cells (or sphincter) at the point where a capillary originates from the metarteriole. This precapillary sphincter can completely stop the blood flow within the capillary.

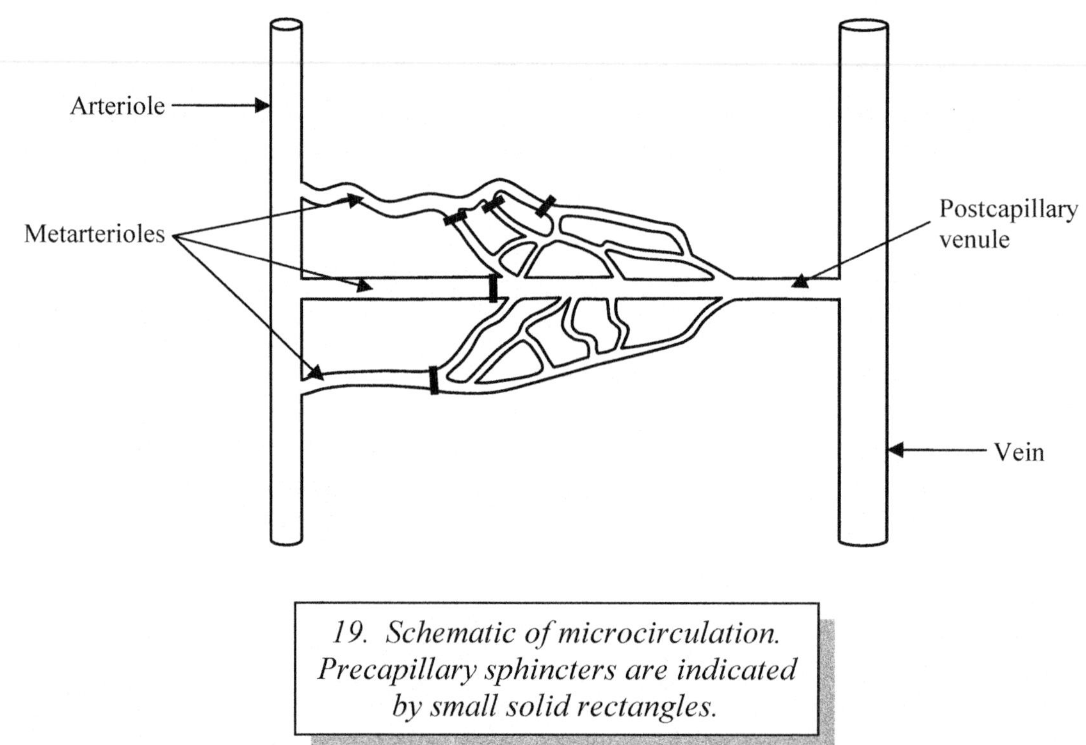

19. Schematic of microcirculation. Precapillary sphincters are indicated by small solid rectangles.

The distribution of capillaries varies with the metabolic activity of the tissue they serve. The muscles, liver, kidneys, and nervous system have extensive capillary networks. Tendons and ligaments have fewer capillaries. All epithelia, the cornea, lens, and cartilage lack capillaries.

Blood from capillaries drains into venules. A venule is a very small tributary of a vein. Just as a river receives water from the many streams that form its tributaries, a vein receives blood from its tributaries.

Veins have the same three layers as arteries, but there are significant differences. The average blood pressure in veins is considerably lower than in arteries. Most of the structural differences between arteries and veins reflect this pressure difference. The tunica interna of veins is thinner. The tunica media of veins is much thinner, with relatively little smooth muscle and elastic fibers. The tunica externa is the thickest layer of veins and consists of collagen and elastic fibers. In addition, many veins inferior to the heart have valves that, in conjunction with contractions of skeletal muscle, facilitate venous return.

Sinusoids and sinuses are unusual types of venous blood channels lacking the tubular shape and some of the structural components of veins. Sinusoids are large, thin-walled, irregularly-shaped venous channels lacking an adventitia but having a lining of reticuloendothelium. They are found in the liver, spleen, lymph nodes, bone marrow, and certain endocrine glands. Sinuses are dilated, irregular venous channels lacking an adventitia, such as the dural venous sinuses that are located between the two layers of dura mater covering the brain (see Topic 24.3).

Varicose veins (varices) are a physiological consequence of human bipedal locomotion.[83] A significant amount of the total blood volume is above the legs in an erect human. This increases gravitational hydrostatic pressure on the venous system in the lower extremities. Consequently, the veins can become dilated, tortuous, twisted, and elongated (varicose) as blood pools in them. The incidence of varicose veins ranges between 10 and 40 percent of persons between 30 and 70 years of age. Severe varicosities can lead to edema and pain and, on rare occasions, to life-threatening blood clots. Two major causes of varicose veins have been suggested: failure of the one-way valves in the veins or failure of the connective tissue in the vein wall. Clearly, the venous system of the human lower limbs is a case of suboptimal design. In contrast, giraffes have modifications to deal with the even greater gravitational hydrostatic pressure on the veins of their lower limbs.[84] As in humans, the blood is returned by means of muscular contractions in the legs, with one-way valves in the veins. However, in giraffes the skin in the feet and lower legs is very stiff and fits like an elastic bandage. This keeps the volume of these regions from increasing to accommodate venous blood, so it can only move upward. Other mechanisms in giraffes that prevent edema include precapillary vasoconstriction and low permeability of capillaries to plasma proteins.

The tissues of the walls of large vessels cannot be nourished by diffusion of nutrients from their lumina and so are supplied by small arteries called vasa vasorum (meaning "vessels of vessels") that are derived either from the main vessel itself or from adjacent arteries. The vasa vasorum give rise to a capillary network within the tunica adventitia that may extend into the tunica media.

The blood vessels in pulmonary and systemic circulations are different in several important ways. Pulmonary arteries have larger diameters, thinner walls, and less elastic tissue than systemic arteries. Consequently, resistance to blood flow is very low and less pressure is needed to move blood through the lungs. Thus, normal pulmonary capillary blood pressure is lower than average systemic capillary blood pressure.

There are several types of circulatory routes through the body. Most capillary networks are arranged in parallel (not serial) circulatory pathways that deliver blood separately to each tissue of the body. In this pattern, the blood flow around the body is: heart → arteries → arterioles → capillaries → venules → veins → heart. Thus, blood passes through only one capillary network before returning to the heart. Some capillary networks are arranged in serial circulatory pathways known as portal systems, in which blood flows through two consecutive capillary networks before returning to the heart. The serial routes include:

- The hepatic portal system, in which venous blood from the digestive tract, pancreas, and spleen passes to the capillaries of the liver before continuing to the heart.
- The hypothalamo-hypophyseal portal system, in which blood passes from the hypothalamus to the capillaries of the adenohypophysis (anterior pituitary gland) before continuing to the heart.
- The glomerular portal system, in which arterial blood from each glomerulus passes to the peritubular (cortical) capillaries and the vasa recta (medullary capillaries) before continuing to the heart.

Anastomoses are another type of blood flow, in which two arteries, two veins, or an artery and a vein merge with each other. There are three types of anastomoses:

- In an arteriovenous anastomosis (shunt), blood flows from an artery directly into a vein when the capillary network is closed. Shunts occur in the fingers, palms, toes, and ears. By reducing blood flow to these exposed surfaces in cold weather, shunts reduce heat loss but thereby increase the risk of frostbite in these areas.
- In an arterial anastomosis, two arteries supplying the same capillary network merge. This provides collateral (alternative) routes of blood supply to the capillary network so that blockage in one artery does not necessarily disrupt blood flow to it. Such blockage can occur when normal movements compress a vessel, or as a result of disease, injury, or surgery. For example, anastomoses between branches of the right and left coronary arteries in the myocardium reduce the risk of heart attack (myocardial infarction). Another arterial anastomosis is the cerebral arterial circle (circle of Willis), formed by anastomoses between branches of the internal carotid and vertebral arteries.
- The most common type of anastomosis is venous, in which two veins from the same capillary network merge, providing alternative routes of drainage. As a result, blockage of a vein is rarely as life-threatening as blockage of an artery. For example, when either the hepatic sinusoids or the hepatic portal vein are obstructed, blood backs up in the portal system. Anastomoses between tributaries

of the hepatic portal system and tributaries of the inferior vena cava then return portal blood to the inferior vena cava.

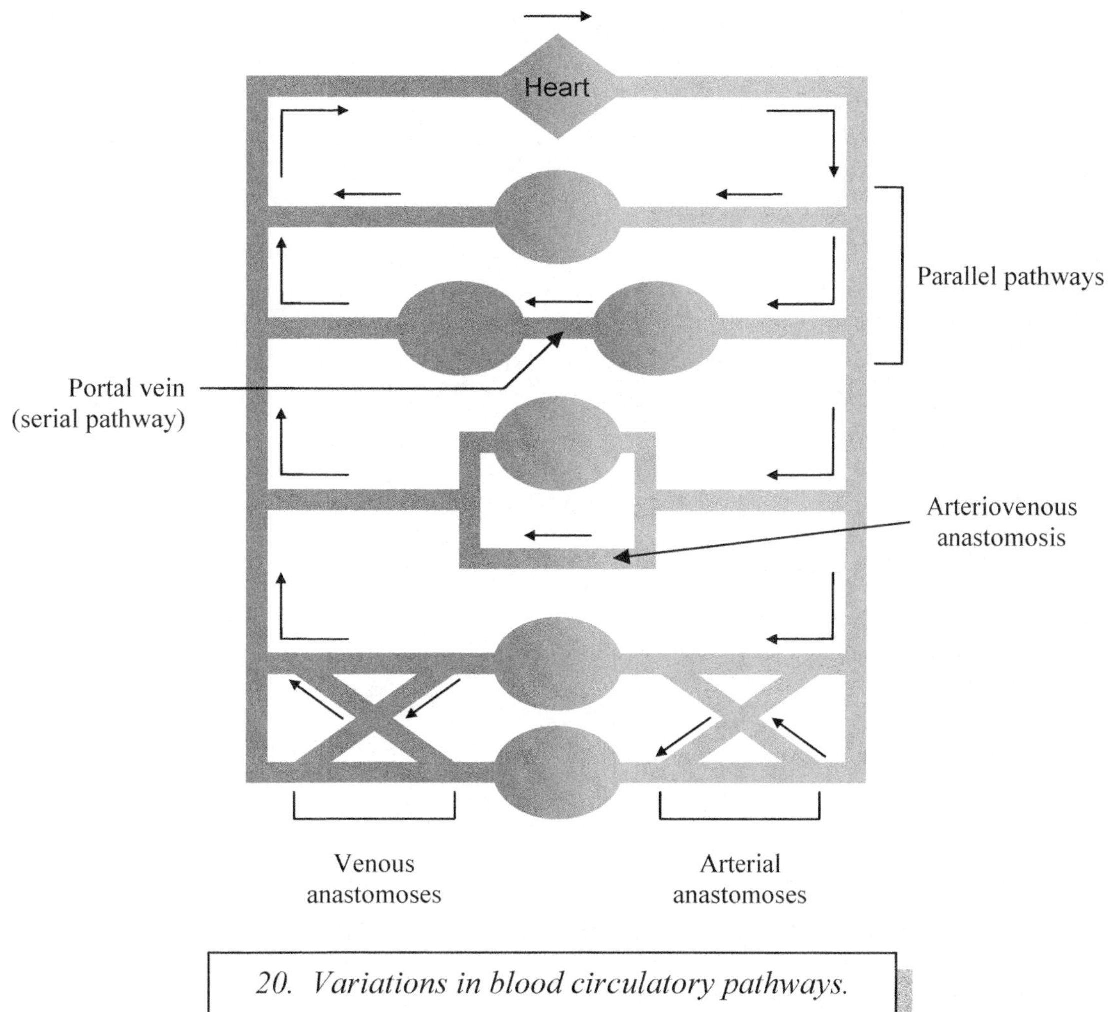

20. Variations in blood circulatory pathways.

Because the lungs, kidneys, and gastrointestinal organs do not begin to function until birth, the circulatory system of a fetus, or fetal circulation, is different from the postnatal circulation. The fetal respiratory organ is the placenta, which permits gas exchange by diffusion between the fetal and maternal bloodstreams while keeping them separate. The umbilical vein transports oxygenated blood from the placenta to the fetus. In the inferior vena cava, blood from the umbilical vein mixes with deoxygenated blood returning from the lower extremities. After entering the right atrium, most of this oxygen-rich blood passes into the left atrium through an opening called the foramen ovale in the interatrial septum. Deoxygenated blood from the superior vena cava also enters the right atrium but is pumped by the right ventricle into the pulmonary trunk. Most of this blood passes into the aorta through the ductus arteriosus, a communication between the pulmonary trunk and aorta. Blood in the left atrium enters the left ventricle, which pumps it into the systemic circulation. Most of the deoxygenated blood in the systemic circulation returns to the placenta by the two umbilical arteries.

When the lungs begin to function at birth, the circulation is converted to the adult pattern. The opening in the interatrial septum between the two atria is closed, leaving a depression called the fossa ovalis, and the ductus arteriosus becomes a small, nonvascular structure called the ligamentum arteriosus.

TOPIC 17.4. THE LYMPHATIC/IMMUNE SYSTEM

As mentioned in Topic 5, the immune system is not really an organ system, but a collection of lymphocytes and macrophages that protect the body against pathogens. Although distributed throughout the body, these cells are closely allied with the lymphatic system. Because the lymphatic system enables the immune system to function efficiently, it is appropriate that they be considered together. The lymphatic/immune system consists of:

- Lymphoid tissue, a specialized form of reticular connective tissue that contains large numbers of lymphocytes and macrophages.
- Lymphoid organs, in which lymphocytes and macrophages are particularly concentrated.
- Lymphatic vessels or lymphatics.
- Lymph, which is tissue fluid that has entered and flows through lymphatic vessels.

The major functions of the lymphatic/immune system are:

- Draining excess tissue fluid and returning it to the blood. More fluid leaves the blood capillaries than is reabsorbed by them. The lymphatic system absorbs this excess tissue fluid and returns it to the bloodstream by way of the lymphatic vessels.
- Transporting dietary lipids. In the small intestine, dietary fat is not absorbed by blood capillaries but by lymphatic vessels called lacteals.
- Protecting against invasion of pathogens. Cells of the immune system have the ability to distinguish **self** (the body's own macromolecules) from **nonself** (foreign substances) and to coordinate the destruction or inactivation of foreign substances that may be isolated molecules or parts of a microorganism. As the lymphatic system collects tissue fluid, it picks up foreign substances that may be present in it and as the fluid is transported back to the cardiovascular system, it passes through dense populations of lymphocytes and macrophages that detect these substances and mount a quick response. Thus, the lymphatic system is the second line of defense, after the epithelial covering of the body.

Lymph is produced as tissue fluid is absorbed into microscopic vessels called lymphatic capillaries located in the spaces between tissue cells. Unlike blood capillaries, lymphatic capillaries are closed at one end. Lymphatic capillaries are located throughout the body except in avascular tissue, the placenta, the central nervous system, portions of the spleen, and red bone marrow. The walls of lymphatic capillaries are constructed so that when hydrostatic pressure is higher in the tissue fluid than in the lymph, fluid and even bacteria and other cells flow into the capillaries. About three liters of excess tissue fluid flow into the lymphatic capillaries each day. When hydrostatic pressure is higher in the lymph than in the tissue fluid, then the walls of the capillaries prevent the passage of fluid. Lymphatic capillaries in intestinal villi collect fat absorbed from the intestine after a meal. If the meal has been particularly fatty, the lymph in these vessels appears milky. This is why these lymphatics are called lacteals and the lymph in them is called chyle.

Lymphatic capillaries merge to form larger lymphatic vessels that merge to form collecting vessels. They resemble veins but have thinner walls and more valves. Traveling close to veins in the subcutaneous layer and close to arteries in the viscera, collecting vessels eventually merge to form larger lymphatic trunks. Each lymphatic trunk, such as the lumbar, intestinal, and intercostal, drains a major portion of the body. The lymphatic trunks converge to form two collecting ducts, the largest of the lymphatic vessels. The thoracic (left lymphatic) duct drains lymph from the left side of the head, neck, and chest, the left upper extremity, and the entire body below the ribs. The right lymphatic duct drains lymph from the upper right side of the body. Each collecting duct drains into the subclavian vein on its respective side.

The lymphatic system has no pump analogous to the heart. The flow of lymph through lymphatic vessels is the result of skeletal muscle contractions and respiratory movements, aided by one-way valves in lymphatic vessels. As a result, physical exercise significantly increases the rate of lymphatic return.

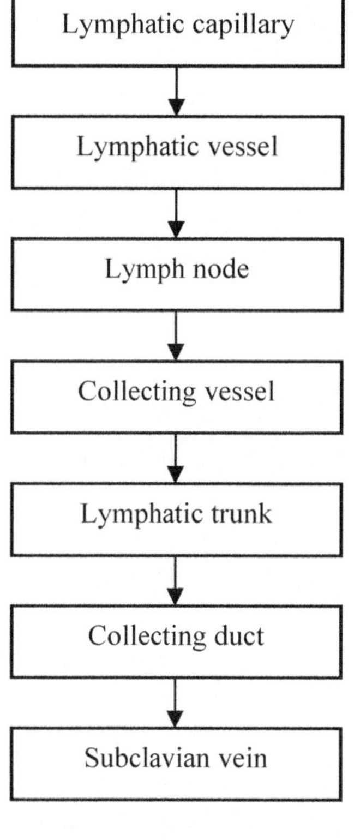

21. The lymphatic pathway.

There are two ways to approach the lymphatic system. Functionally, it can be divided into two parts. Red bone marrow contains stem cells that give rise to mature B lymphocytes and to immature T lymphocytes that migrate to the thymus where they mature. For this reason, the bone marrow and the thymus are called primary lymphoid organs. Lymphocytes produced by the primary lymphoid organs colonize secondary or peripheral areas of the body where lymphoid tissues occur in various forms. These secondary lymphoid organs and tissues, which include diffuse lymphoid tissue, the lymph nodes, lymphoid nodules, tonsils, and spleen, are the sites where most immune responses occur. Structurally, the lymphatic system can be divided into three parts: diffuse lymphoid tissue; dense, nonencapsulated lymphoid tissue; and dense encapsulated lymphoid tissue (the lymphoid organs).

Diffuse lymphoid tissue is the simplest form of lymphoid tissue, consisting of lymphocytes scattered throughout the mucous membranes and connective tissues of many organs. When associated with the mucous membranes of the respiratory, digestive, urinary, and reproductive tracts, it is called mucosa-associated lymphoid tissue (MALT); diffuse lymphoid tissue associated with just the digestive tract is called gut-associated lymphoid tissue (GALT).

Dense, nonencapsulated lymphoid tissue consists of lymphocytes clustered in dense oval masses called lymphoid nodules. The nodules are usually small, solitary, discrete, and transient. In the small intestine at its junction with the large intestine, lymphoid nodules are a relatively constant feature. Here, the nodules form clusters called Peyer's patches. Large aggregations of lymphoid nodules form the five tonsils. Two lingual tonsils are located in the root of the tongue, two palatine tonsils are located in the lateral, posterior regions of the oral cavity, and one medial pharyngeal tonsil is located on the posterior wall of the nasopharynx. The palatine tonsils are the largest and most often infected. A hypertrophied pharyngeal tonsil is called an adenoid.

Dense, encapsulated lymphoid tissue, or the lymphoid organs, includes the thymus, lymph nodes, and the spleen. The thymus is where T-lymphocytes mature. It is located in the superior mediastinum, between the sternum and aortic arch. It usually has two lobes, each enclosed by a capsule of connective tissue. The thymus is both a lymphoid and an endocrine organ. Lymph nodes are encapsulated, oval structures located along lymphatic vessels. Lymph enters a node through several afferent lymphatic vessels, flows through sinuses (wide channels), and then exits the node through one or two efferent lymphatic vessels. As lymph flows through a lymph node:

- Foreign substances are filtered and trapped by reticular fibers.
- Macrophages destroy these foreign substances by phagocytosis.
- Lymphocytes destroy other foreign substances by means of immune responses.
- Plasma cells and T-lymphocytes can enter the lymph and circulate to other parts of the body.

The spleen is the largest single mass of lymphoid tissue in the body. It is located in the left hypochondriac region between the stomach and thoracic diaphragm. It is enclosed by a capsule of dense connective tissue. It is filled primarily with red pulp, a highly vascular tissue where phagocytosis of bacteria and worn-out erythrocytes occurs. Scattered throughout the red pulp is the white pulp, lymphoid tissue where B-lymphocytes proliferate into plasma cells.

TOPIC 18. BIPEDALISM, BIG BRAINS, AND BARE SKIN

The American primatologist and anthropologist Sherwood L. Washburn (1911-2000) first pointed out that during the recent evolutionary history of hominids, three regions of the body evolved somewhat independently, that is, the body evolved in a mosaic fashion:

- The thorax and shoulder. Humans share the anatomical features of these regions with the great apes and gibbons because their common ancestor was a suspensory feeder, adapted to hanging and swinging under tree branches. This anatomical complex includes mobile shoulder joints, long arms, and a short lumbar region. The upper limbs of humans, however, are shorter and less massive than those of modern apes.

- The pelvis and lower limbs. The short, broad shape of the ilium and the altered attachment points for gluteal muscles appeared in the australopithecines and confirmed their pattern of bipedal locomotion. The feet changed shape in response to being the body's only support during locomotion (see Topic 15).
- The skull. The cranium of humans increased dramatically in size, accommodating a much larger brain. At the same time, the human mandible and dentition decreased in size due to changes in diet. As a result, the face is more orthognathic (vertical) in modern humans.

The fossil record shows that after humans diverged from the other apes, evolution in the pelvis and lower limbs occurred first, followed by evolution in the skull.[85] The most recent common ancestor of humans and the other apes lived about 5 or 6 million years ago. About 4 million years ago, bipedal locomotion became established in the australopithecines. At least 2 million years ago, cranial expansion and dental reduction began. The human brain today averages three times the size one would expect for a nonhuman primate of equivalent body size.

However, brain tissue has very high energy demands per unit weight, roughly 16 times greater than those of muscle tissue. The modern adult human body allocates about 20 to 25 percent of its daily energy budget to brain metabolism. In contrast, nonhuman primates typically allocate only 8 to 10 percent, and nonprimate mammals only 3 to 5 percent, to the brain. As a result, the brain generates significantly more heat than other tissues and is particularly sensitive to changes in body temperature. A change of only 4° C begins to disturb the functions of the contemporary human brain. For example, high fevers in children can cause convulsions, indicating abnormal functioning of the nerve cells of the overheated brain.

Since early humans probably originated in the equatorial African savanna habitat, keeping the body and brain cool from heat absorbed from the sun and heat generated by brain metabolism and muscular action must have been essential for survival. Thus, several adaptations evolved for thermoregulation.

One adaptation that evolved was "bare" skin. Humans, chimpanzees, and gorillas differ from other primates in having eccrine sudoriferous (sweat) glands, but humans have the most. Humans have about two million such glands, distributed on the whole body and concentrated on the trunk. Eccrine sudoriferous glands respond to heat stress by producing sweat that cools the skin by evaporation. Heat is further dissipated when blood vessels in the skin dilate and give off heat to the ambient air. Hair reduces the effectiveness of both of these processes, so reduction in hair shaft length on most regions of the body is part of this cooling adaptation. Humans have just as many hair follicles as their shaggy ape relatives, but the hair is much finer and gives the human body a naked and hairless look. However, the loss of thick hair makes the skin vulnerable to the damaging effects of ultraviolet light, so deeply pigmented skin evolved in early humans for protection in equatorial Africa. Skin color seems to make no difference for survival in apes, but is vital for humans.

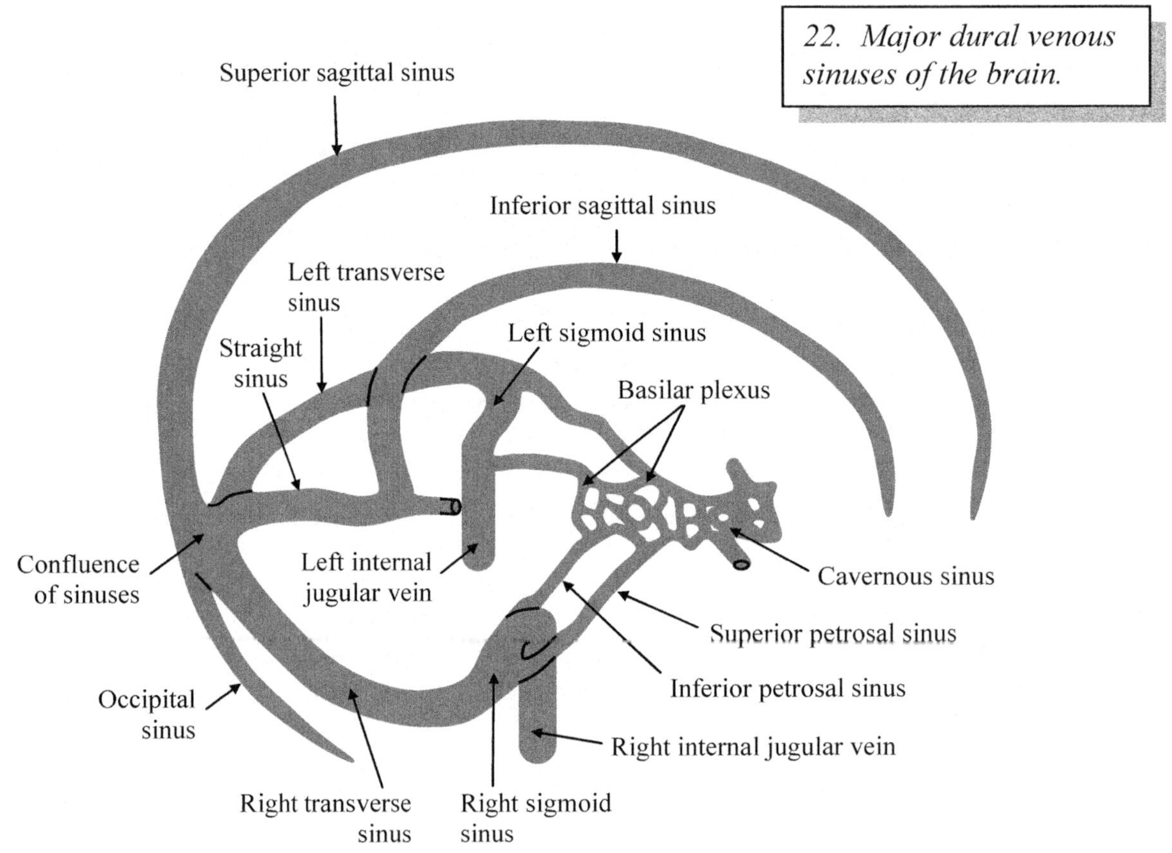

22. *Major dural venous sinuses of the brain.*

Another structure that may have evolved, at least in part, as an adaptation in early humans for thermoregulation is the cranial blood drainage system. Although blood in the scalp and face can return to the heart directly through the external jugular vein outside the skull, it can also follow a more circuitous path by entering the braincase and ultimately exiting through the internal jugular vein. In this path, emissary veins and the cavernous sinus connect scalp veins and facial veins, respectively, to dural venous sinuses

Table 7. The Cranial Blood Drainage System

Name	Drains to
Inferior sagittal sinus	Straight sinus
Superior sagittal sinus	Confluence of sinuses
Straight sinus	Confluence of sinuses
Occipital sinus	Confluence of sinuses
Confluence of sinuses	Transverse sinuses
Cavernous sinus	Superior and inferior petrosal sinuses
Transverse sinuses	Sigmoid sinuses
Superior petrosal sinuses	Sigmoid sinuses
Inferior petrosal sinuses	Internal jugular veins
Sigmoid sinuses	Internal jugular veins

(see Topic 24.3) surrounding the brain. Emissary veins, such as the parietal and mastoid, enter the skull through holes called emissary foramina.

When humans are experiencing hyperthermia (overheating), such as during intense exercise, the face becomes flushed as arteries dilate to bring more blood near the skin surface, which is cooled by the evaporation of sweat. Much of this extra blood, now cooled, enters the emissary veins and cavernous sinus and is delivered into the meningeal veins and sinuses of the dura mater, which covers the brain. From there, some of the blood flows to veins within the surface of the brain itself. As it circulates, the blood removes heat from all these regions.[86] By the time it joins other venous blood leaving the skull, it is warmer than the oxygenated, arterial blood that supplies the brain. This venous network does not have valves, allowing the blood to move freely as required. The frequencies of the parietal and mastoid emissary foramina increased dramatically as brain size increased from *Australopithecus africanus* to *Homo sapiens*,[87] supporting the "radiator hypothesis." This hypothesis posits that evolution of the emissary veins, which are not found in other apes, released a thermal constraint that previously kept brain size within ape ranges.

Interestingly, since the cavernous sinus receives venous blood from the facial veins (via the superior and inferior ophthalmic veins), infections of the face including the nose, tonsils, and orbits can spread easily to it by this route. The internal carotid artery with its surrounding sympathetic plexus passes through the cavernous sinus. The third, fourth, and sixth cranial nerves are attached to the lateral wall of the sinus. The ophthalmic and maxillary divisions of the fifth cranial nerve are embedded in the wall. As a result, an infection in the cavernous sinus that causes a blood clot, a condition called cavernous sinus thrombosis or CST, is life-threatening. Typically, death is due to sepsis or central nervous system infection. Before antibiotics were discovered, the mortality from CST was 80% to 100%; since the discovery of antibiotics, the mortality ranges between 20% and 30%. Morbidity, however, remains high, and complete recovery is rare. Roughly one sixth of patients are left with some degree of visual impairment, and one half have cranial nerve deficits. Fortunately, the incidence of CST is very low. However, this intimate juxtaposition of veins, arteries, nerves, meninges, and paranasal sinuses is clearly **not** the product of Intelligent Design.

TOPIC 19. AN OVERVIEW OF THE DIGESTIVE SYSTEM

Digestion is the breaking down of larger food molecules into molecules that are small enough to enter body cells. Absorption is the passage of these small molecules into blood and lymph. The organs that collectively perform these functions comprise the digestive system. The digestive or gastrointestinal (GI) tract, or alimentary canal, is a tube extending from the mouth to the anus through the ventral body cavity. It is divided into the oral (buccal) cavity, pharynx, esophagus, stomach, small intestine, and large intestine. The accessory digestive organs include the teeth, tongue, salivary glands, pancreas, liver, and gallbladder. The functions of the digestive system include:

- Ingestion (eating).

- Secretion of water, buffers, and digestive enzymes.
- Movement of food through the alimentary canal.
- Digestion by mechanical processes. This consists of the chewing (mastication) of food by the teeth to break it into pieces small enough to swallow and the churning of food by the smooth muscles of the stomach and small intestine to thoroughly mix it with digestive enzymes.
- Digestion by chemical processes. This consists of a series of catabolic reactions catalyzed by enzymes in which large food molecules (carbohydrates, lipids, protein, and nucleic acids) are broken down into smaller molecules that may be absorbed and used by body cells.
- Absorption into the blood and lymph.
- Defecation of indigestible substances and bacteria.

The wall of the alimentary canal, especially from the esophagus to the anus, consists of four major layers (tunics), which from deep to superficial are: the mucosa, the submucosa, the muscularis, and the adventitia. The adventitia in some places is replaced by the serosa.

The mucosa is a mucous membrane that surrounds the lumen and consists of:

- An inner lining layer of epithelium. Nonkeratinized stratified squamous epithelium occurs in the oral cavity, pharynx, esophagus, and anal canal for protection. Simple columnar epithelium occurs throughout the rest of the alimentary canal for secretion and absorption.
- A middle layer of areolar connective tissue called the lamina propria. It contains blood vessels, lymphatic vessels, and scattered lymphoid nodules as well as most components of the mucosa-associated lymphoid tissue (MALT).
- An outer layer of smooth muscle tissue called the muscularis mucosae. It creates small folds in the stomach and intestinal mucosa that increase the surface area for digestion and absorption.

The submucosa is composed of areolar connective tissue with many blood and lymph vessels and a submucosal (or Meissner's) nerve plexus. It may also contain glands and lymphoid tissue.

The muscularis consists of muscle tissue. Skeletal muscle tissue occurs in the oral cavity, pharynx, upper portion of the esophagus, and the external anal sphincter. Smooth muscle tissue occurs in the rest of the alimentary canal. The muscularis with smooth muscle is usually divided into two sublayers according to the orientation of the muscle cells: an inner circular sublayer and an outer longitudinal sublayer. Between the two sublayers lies the myenteric (or Auerbach's) nerve plexus as well as connective tissue with blood and lymph vessels. Contractions of the muscularis help break down food physically, mix it with digestive secretions, and propel it along the alimentary canal.

The adventitia is found above the diaphragm and serosa below the diaphragm, where it is called the visceral peritoneum. The serosa is a serous membrane composed of a thin layer of areolar connective tissue covered by a layer of simple squamous epithelium called mesothelium.

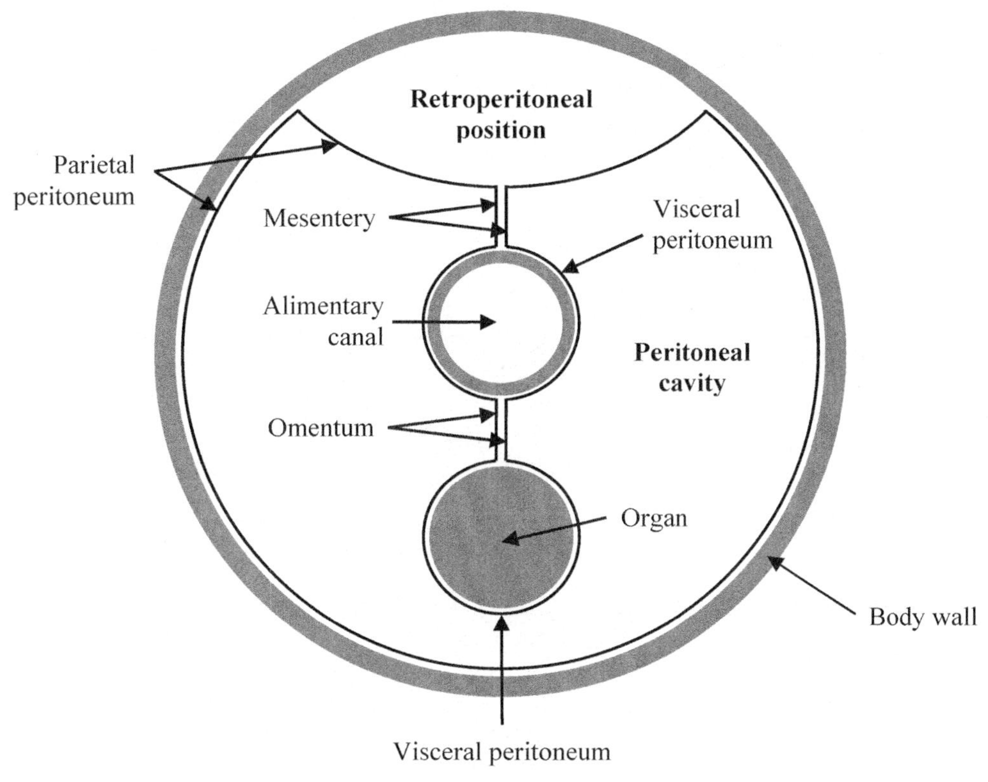

23. Schematic diagram showing relationship of parietal to visceral peritoneum.

The submucosal nerve plexus and the myenteric nerve plexus are part of a nervous network called the enteric nervous system. It is located in the esophagus, stomach, and intestines and regulates their motility, secretion, and blood flow.

The peritoneum is the largest serous membrane in the body. The parietal peritoneum, which lines the wall of the abdominal cavity, and the visceral peritoneum (serosa), which covers some of the organs of the abdominal cavity, is actually one continuous membrane. The narrow gap between the two membranes, called the peritoneal cavity, contains serous fluid. This space can fill with several liters of fluid in disease and trauma. Organs lying between the parietal peritoneum and the posterior abdominal wall, such as the kidney and duodenum, are said to be retroperitoneal.

Parts of the peritoneum are arranged in large folds. These folds function to bind organs to each other and to the walls of the abdominal cavity. They also contain blood vessels, lymphatic vessels, and nerves that supply the abdominal organs. Mesenteries are peritoneal folds suspending organs from the body wall, such as the mesocolon and the common mesentery. The mesocolon attaches the transverse and sigmoid colon to the posterior abdominal wall, and the common mesentery attaches the jejunum and ileum to the same wall. Omenta (singular: omentum), or ligaments, are peritoneal folds suspending one organ from another organ. The falciform ligament attaches the liver to the anterior abdominal wall and thoracic diaphragm. The lesser omentum suspends the stomach from the liver. The greater omentum suspends the transverse colon from the stomach, contains large quantities of adipose tissue, and is the largest peritoneal fold.

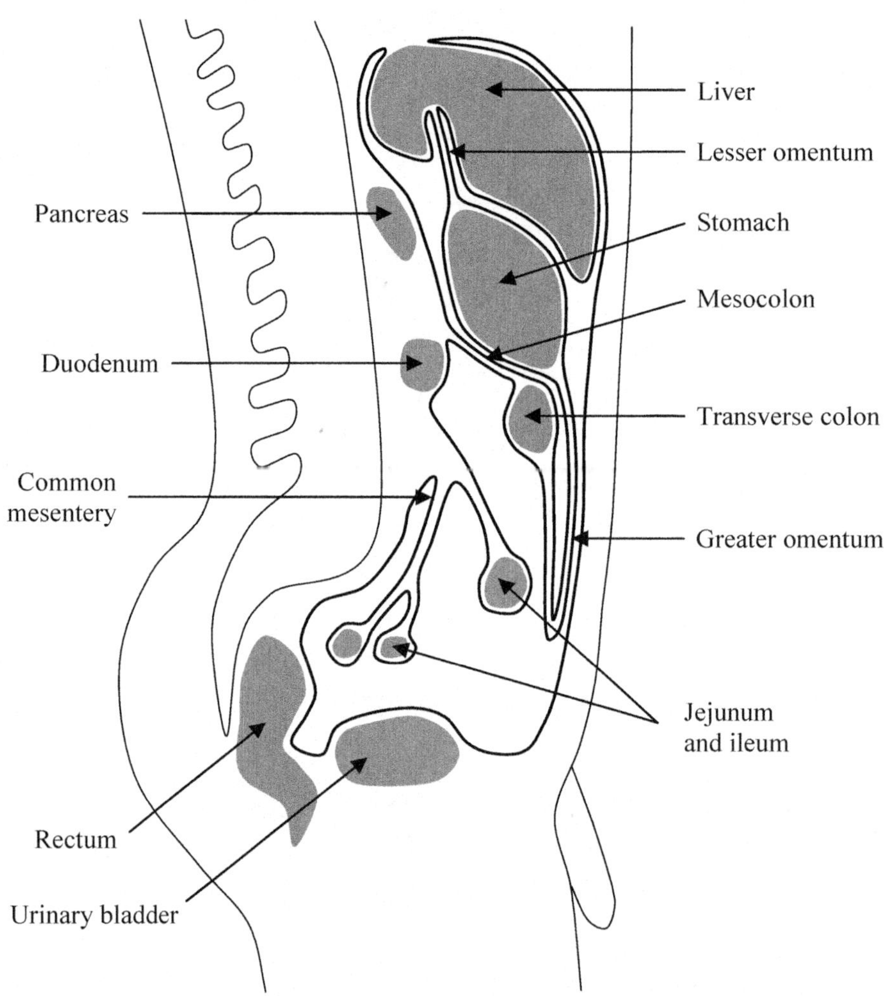

24. Schematic diagram of peritoneum of the abdominal cavity. Midsagittal section, right lateral view.

TOPIC 19.1. THE ORAL CAVITY AND ACCESSORY ORGANS

The oral (or buccal) cavity is formed by the cheeks, hard and soft palates, and tongue. The cheeks consist of skeletal muscle covered externally with skin and internally with mucous membrane lined by nonkeratinized stratified squamous epithelium. The anterior opening of the oral cavity is called the oral orifice (mouth) and is surrounded by lips (labia). The posterior opening to the oropharynx is called the fauces. The hard palate, which forms the anterior roof of the oral cavity, consists of the maxillary and palatine bones. The soft palate (velum palatinum), which forms the posterior roof of the oral cavity, is a muscular structure lined by mucous membrane. The uvula is a muscular process that hangs from the free border of the soft palate. During swallowing the soft palate and uvula are elevated, blocking the nasopharynx and preventing regurgitation into the nasal cavity.

The tongue is an accessory digestive organ composed of skeletal muscles covered by mucous membrane. The posterior one-third of the tongue, called the root, is demarcated from the anterior two-thirds, called the body, by a **V**-shaped groove called the sulcus terminalis. The body of the tongue occupies the oral cavity and the root occupies the oropharynx. The body is attached to the floor of the oral cavity by a fold of mucous membrane called the lingual frenulum. It limits movement of the tongue posteriorly. Extrinsic muscles originate outside the tongue and insert into it. They move the tongue for food manipulation during chewing and swallowing. Intrinsic muscles originate and insert within the tongue. They alter the shape and size of the tongue for speech and swallowing. The mucosa of the dorsal surface of the body of the tongue is formed into lingual papillae, which are projections of the lamina propria covered with epithelium. There are three types:

- Filiform papillae, which lack taste buds. They are the most common papillae, appearing as whitish, cone-shaped projections distributed in parallel rows.
- Fungiform papillae, most of which contain taste buds. They are red mushroom-like projections distributed among the filiform papillae and are more numerous near the tip of the tongue.
- Circumvallate papillae, all of which contain taste buds. These papillae are arranged in a row immediately anterior to the sulcus terminalis of the tongue and are surrounded by ducts of lingual glands that secrete lingual lipase and mucus.

The three pairs of salivary glands secrete major quantities of saliva when food enters the oral cavity. Two parotid glands are located inferior and anterior to the ears, between the skin and masseter muscle. Two submandibular glands are located beneath the base of the tongue. Two sublingual glands are located in the floor of the oral cavity superior to the submandibular glands. Saliva is 99.5 percent water and 0.5 percent solutes. The latter include: lysozyme, a bacteriolytic enzyme; salivary amylase, which initiates starch digestion; and lingual lipase, which initiates triglyceride digestion.

The teeth, or dentes, are embedded in mandibular and maxillary sockets called dental alveoli, forming joints called gomphoses. Each socket is lined by a periodontal ligament, also called a periodontium, which is a modified periosteum that anchors the tooth in position and acts as a shock absorber during mastication. The gum, or gingiva, covers the alveolar bone. A typical tooth consists of three major regions defined by their relationship to the gingiva. The crown is the visible portion located above the level of the gums. One to three roots are embedded in the socket. The neck is the narrow junction of the crown and roots near the gum line.

A tooth is composed of several substances. The crown contains a cavity connected to a canal in the root, both of which are filled with dental pulp. Pulp is composed of loose connective tissue, blood and lymphatic vessels, and nerves. However, most of the tooth is made of dentin, which is a living calcified connective tissue. The dentin of the crown is covered by enamel produced before the tooth erupts. Enamel is not a tissue but a noncellular secretion which is composed of calcium salts and is the hardest substance in the body. Enamel protects the tooth against the wear of chewing and is a barrier to acids that may easily dissolve the dentin. However, because it is nonliving, damaged enamel cannot regenerate but must be artificially repaired. The dentin of the root is covered by cementum, a bonelike substance that attaches the root to the periodontal ligament. Like dentin, cementum is a living connective tissue that can regenerate if damaged.

Human teeth are typical of mammals' teeth, which are specialized into a number of types, a feature known as heterodonty. Each half of each jaw usually contains:

- Two incisors, which are chisel-like and used for cutting off pieces of food.
- One cuspid (canine), which is used to tear and shred food but (unlike most mammals) does not project beyond the level of the other teeth. The size of the roots of the upper cuspids is another fact of the human body that cannot be explained without evolution. In monkeys, the cuspids are much larger than they are in humans, and as such, they require much larger roots to anchor them. In humans the cuspids are now much smaller, but if you run your finger over your upper gums, you can feel the bumps of your unnecessarily large cuspid roots (even through your lip).
- The cheek teeth, consisting of two premolars, or bicuspids, and three molars. They are used to crush and grind food.

There are two sets of teeth, or dentitions. The deciduous dentition, consisting of 20 teeth, begins to erupt around 6 months after birth. The permanent dentition, consisting of 32 teeth, replaces the deciduous and begins to erupt around 6 years of age. The last teeth in humans to develop and erupt into the jaws are the third molars, nicknamed the "wisdom teeth" since their appearance coincides with passage into adulthood around age 18.

Like the first and second molars, the third molars evolved in early humans to grind tough, raw food to a texture that made it safe to swallow.[88] Early humans had much less brain space and a larger, more powerful jaw than modern humans. The three molars created a large, effective chewing table that suited their diet and lifestyle. Natural selection apparently favored a larger brain, so the brain cavity expanded while the jaws diminished accordingly.[89] However, the number of teeth in the normal human jaw remained the same (32) throughout this change. As a result, the third molars are now usually trying to erupt into a jaw that is too small (unless a few teeth have already been lost due to decay), which often leads to problems. When a third molar does not fully erupt through the gums, it cannot be cleaned properly and can collect food debris, bacteria and plaque around itself. This can result in tooth decay, gum disease, infection and abscess of not only the third molar, but of the nearby molars and surrounding gum tissue as well. Teeth that are impacted (fail to erupt) can lead to cyst formation and other destructive pathology. These conditions may be accompanied by pain and swelling, or no symptoms at all, even though the other teeth in the oral cavity may be at risk of damage. All of the above problems often necessitate surgical removal of third molars, with no adverse consequences; therefore, third molars are considered to be vestigial in humans.

TOPIC 19.2. THE PHARYNX

Mastication (chewing) in the oral cavity converts the food to a soft, flexible mass called a bolus. Mastication is the first step in mechanical digestion and requires little thought because food stimulates receptors that trigger an involuntary chewing reflex.

In swallowing, or deglutition, the bolus of food first enters the pharynx. The pharynx is a funnel-shaped tube lined by mucous membrane, extending from the choanae to the esophagus posteriorly and the larynx anteriorly. It has a deep layer of longitudinally oriented skeletal muscle and a superficial layer of circular skeletal muscle. Deglutition is a complex action involving over 22 muscles in the mouth, pharynx, and esophagus. They are coordinated by the swallowing center, a nucleus in the medulla oblongata and pons of the brain.

TOPIC 19.3. THE ESOPHAGUS

The esophagus is a straight muscular, collapsible tube 25 to 30 cm long. It travels from the laryngopharynx downward through the mediastinum of the thorax, anterior to the spine and posterior to the trachea. It passes through the esophageal hiatus in the thoracic diaphragm, continues another 3 to 4 cm, and connects to the stomach at an opening called the cardiac orifice (named for its proximity to the heart). There are several distinctive characteristics of the wall of the esophagus:

- Mucous glands are found in the mucosa of the esophagus near the stomach and in the submucosa of the entire esophagus.
- The muscularis in the superior third of the esophagus is skeletal muscle, in the middle third both skeletal and smooth, and in the inferior third entirely smooth muscle.
- The outermost layer above the diaphragm is adventitia, consisting of areolar connective tissue. Its connective tissue merges with the connective tissue of the structures surrounding the esophagus. The outermost layer of the short section below the diaphragm is a serosa.

The functions of the esophagus include secreting mucus and transporting food to the stomach. The mucus acts as a lubricant to facilitate movement of the bolus through the esophagus. The entry of food into the esophagus is regulated by the upper esophageal sphincter (UES), which opens in the process of swallowing. Food is pushed through the esophagus by involuntary waves of muscular contraction called peristalsis, which is a function of the muscularis. Just above the thoracic diaphragm is the lower esophageal sphincter (LES) or gastroesophageal sphincter, which is a physiological sphincter rather than an anatomical sphincter because there is no thickening of the muscularis. The LES briefly relaxes to permit passage of the bolus from the esophagus into the stomach; it normally prevents reflux of acid from the stomach into the esophagus.

TOPIC 19.4. THE STOMACH

The stomach is a **J**-shaped, mixed exocrine-endocrine organ located directly under the thoracic diaphragm in the upper left portion of the abdominal cavity. Its medial margin is called the lesser curvature and its longer lateral margin is called the greater curvature. The stomach functions primarily as a food storage organ, with most digestion occurring after the food passes on to the small intestine. It consists of four major areas:

- The cardiac region, or cardia, surrounds the cardiac orifice.
- The fundic region, or fundus, is the rounded portion above and to the left of the cardiac region.
- The body, or corpus, is the large central portion that makes up the greatest part of the stomach inferior to the cardiac orifice. The fundic region and the body are identical in microscopic structure.
- The pyloric region is the lower portion that connects to the duodenum. It consists of the funnel-like pyloric antrum, which connects to the body of the stomach, and the narrower pyloric canal. The pyloric canal terminates at the pylorus, which is the opening to the duodenum.

When the stomach is empty, its mucosa lies in large folds called rugae. The mucosa of the stomach wall is specialized for secretion. Its surface consists of simple columnar epithelial cells called mucous surface cells. Invaginations of the surface epithelium into the lamina propria form gastric pits. Several tiny glands open into the bottom of each gastric pit. The glands in the cardiac and pyloric regions are called cardiac and pyloric glands, respectively, and secrete mostly mucus. The glands in the rest of the stomach are called gastric glands, which have a greater variety of cell types, producing several secretions. The upper half, or neck, of each gastric gland consists of stem, parietal, and mucous neck cells, while the lower half, or base, contains parietal, chief, and enteroendocrine cells:

- Stem cells are present but infrequent in the upper region of the neck. They divide rapidly, with some daughter cells moving upward to replace the pit and surface mucous cells and others moving downward to replace mucous neck cells and parietal, chief, and enteroendocrine cells.
- Mucous neck cells secrete mucus.
- Parietal cells, which are present mostly in the neck, secrete hydrochloric acid

(HCl) and intrinsic factor. Intrinsic factor is required for absorption of vitamin B_{12}.

- Chief cells are the most common cells in the base of the gastric glands. They secrete pepsinogen and gastric lipase.
- Enteroendocrine cells located primarily in the lower region of the base secrete hormones that regulate digestion. For example, G cells in the pyloric glands secrete gastrin, which stimulates the secretion of acid by the parietal cells of gastric glands.

The exocrine secretions of the mucous, chief, and parietal cells form gastric juice, which flows into the stomach lumen. In the lumen, the low pH converts pepsinogen into pepsin. Pepsin is an enzyme that functions to break dietary proteins into shorter peptide chains, which then pass to the small intestine where their digestion is completed. Gastric lipase performs limited digestion of triglycerides.

The muscularis of the stomach wall has three layers of smooth muscle tissue to permit contractions in a variety of directions. The outer layer is longitudinal, the middle layer is circular, and the inner layer is oblique. The oblique layer is found primarily in the body of the stomach. Contractions of the muscularis churn and break apart the food and mix it with gastric juice, forming a thin, acidic liquid called chyme. At the pylorus, the circular layer is greatly thickened, forming the pyloric (gastroduodenal) sphincter. Only small amounts of chyme are allowed into the duodenum at a time, enabling the duodenum to neutralize the stomach acid and digest nutrients little by little. Depending upon the chemical composition of a meal, the stomach empties all its contents into the duodenum about two to six hours after ingestion.

The stomach does not absorb any significant amount of nutrients but does absorb aspirin and some lipid-soluble drugs. It also absorbs electrolytes and some water. Ethanol (grain alcohol) is absorbed mainly by the small intestine, so its intoxicating effect depends partly on how rapidly the stomach is emptied.

TOPIC 19.5. THE PANCREAS

The pancreas is a soft, spongy, pink organ located posterior to the greater curvature of the stomach and connected, usually by two ducts, to the duodenum. It consists of a medial head encircled by the duodenum, a middle portion called the body, and a lateral, tapered tail.

The pancreas is both an endocrine and exocrine gland. Its endocrine part contains about one percent of the glandular epithelial cells, which form clusters called pancreatic islets (islets of Langerhans) that secrete the hormones glucagon, insulin, somatostatin, and pancreatic polypeptide. Its exocrine part contains the remaining 99 percent of the glandular epithelial cells, which form clusters called acini (singular: acinus) that secrete pancreatic juice.

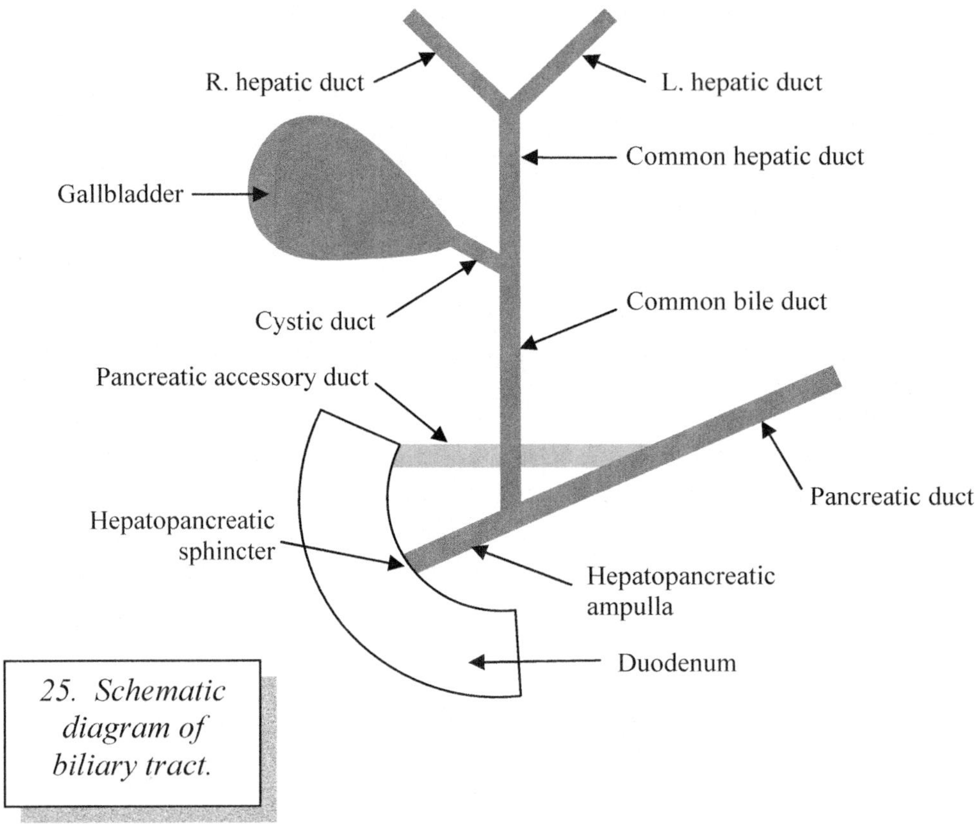

25. Schematic diagram of biliary tract.

Pancreatic juice is an alkaline mixture (pH = 8) of water, electrolytes, and digestive enzymes. The main electrolyte is sodium bicarbonate, which buffers hydrochloric acid from the stomach, stops the action of pepsin, and provides the proper pH for digestive enzymes in the small intestine to function. The digestive enzymes that pancreatic juice contributes to the duodenum include pancreatic amylase, trypsin, chymotrypsin, carboxypeptidase, elastase, pancreatic lipase, ribonuclease, and deoxyribonuclease. Pancreatic juice flows from the acini into small ducts that eventually converge on the pancreatic duct (duct of Wirsung).

The pancreatic duct runs lengthwise through the middle of the pancreas and joins the common bile duct from the liver and gallbladder at an expanded chamber called the hepatopancreatic ampulla (ampulla of Vater). At the passage between the ampulla and the duodenum is the hepatopancreatic sphincter (sphincter of Oddi), which regulates the release of both bile and pancreatic juice into the duodenum. However, in many people, a smaller, pancreatic accessory duct (duct of Santorini) branches from the pancreatic duct and empties directly into the duodenum, allowing pancreatic juice to bypass the hepatopancreatic sphincter.

TOPIC 19.6. THE LIVER AND GALLBLADDER

The liver has numerous functions, only one of which, the secretion of bile, contributes to digestion. Located immediately inferior to the diaphragm in the upper portion of the abdominal cavity, the liver is the heaviest gland in the body and, after the skin, is the second largest organ in the body. It consists of two major lobes, of which the right is larger than the left, and two minor lobes, the quadrate and the caudate. The major lobes are separated by the falciform ligament, which is a fold of the parietal peritoneum that suspends the liver from the diaphragm and anterior abdominal wall. The liver is almost completely covered by serosa (visceral peritoneum).

The lobes of the liver consist of functional units called lobules, which are composed of specialized epithelial cells called hepatocytes, or hepatic cells. There are about one million liver lobules, each about 1 mm in diameter and 2 mm long. The lobules are hexagonal in cross section, with sinusoids (wide capillaries) radiating from a central vein. The areas between the sinusoids consist of only one or two layers of hepatocytes, called hepatic plates; this arrangement allows each cell to be in close contact with the blood. The sinusoids are lined by endothelium and partly by phagocytes called stellate reticuloendothelial (or Kuppfer's) cells. The latter cells destroy bacteria, toxins, and worn-out leukocytes and erythrocytes. At each of several different corners of the hexagon is a portal (hepatic) triad, which consists of a bile ductule, hepatic portal vein, and hepatic artery.

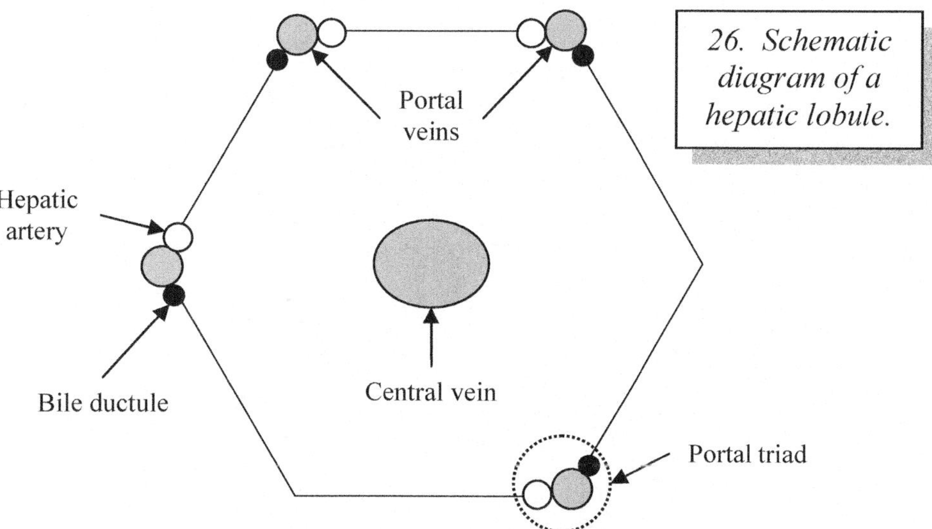

26. *Schematic diagram of a hepatic lobule.*

The hepatocytes continually absorb some substances and secrete others. Deoxygenated hepatic portal venous blood and oxygenated arterial blood mix and flow from the portal triads through the sinusoids of the lobule and into the central vein. As they remove certain poisons, the hepatocytes receive oxygen and nutrients. The central veins converge to eventually form two hepatic veins that carry blood from the liver directly to the inferior vena cava. At the same time, hepatocytes secrete bile into thin channels called bile canaliculi (sing.: canaliculus) located within each hepatic plate. Bile flows through the canaliculi to the bile ductule located within each the portal triad. As a result of the structure of the lobule, blood and bile never mix in the liver under normal conditions. The bile ductules converge to eventually form the right and left hepatic ducts, which join to form the common hepatic duct that exits the liver.

Bile is partially an excretory product and partially a digestive secretion. It contains several important substances, including bile salts and bile pigments. Bile salts are certain steroids synthesized from cholesterol. Their secretion is the only function of the liver involved with digestion. All other components of the bile are waste, to be excreted by way of the intestines. Bile salts are not excreted in the feces but are reabsorbed in the ileum, returned to the liver, and resecreted. In the small intestine, they aid in emulsification (dissolving) of fats and in absorption of the products of fat digestion. Bile pigments consist of biliverdin and bilirubin. Biliverdin is a greenish pigment produced by the decomposition of hemoglobin. Most of the biliverdin is converted to bilirubin, a yellow-green pigment. Bilirubin is broken down by bacteria in the small intestine, producing urobilinogen that gives feces their normal brown color. In the absence of bile secretion, the feces are grayish white and marked with streaks of undigested fat.

The common hepatic duct merges with the cystic duct from the gallbladder, forming the common bile duct. The common bile duct and the pancreatic duct merge to form the hepatopancreatic ampulla, which empties into the duodenum. When the small intestine is empty, the hepatopancreatic sphincter closes. This causes bile to accumulate in the common bile duct and enter the cystic duct and the gallbladder, where it is stored.

The gallbladder is a pear-shaped, greenish sac that is located in a fossa of the visceral surface of the right lobe of the liver. It absorbs water and ions from the bile, which it stores in this concentrated state until it is needed in the small intestine. The wall of the gallbladder has a highly folded mucosa with simple columnar epithelium, but there is no submucosa. The muscularis, composed of smooth muscle tissue, contracts upon hormonal stimulation to eject stored bile into the cystic duct. The outer layer of the wall is a serosa, part of the visceral peritoneum.

TOPIC 19.7. THE SMALL INTESTINE

Nearly all chemical digestion and nutrient absorption occur in the small intestine, including 90 percent of all water absorption. The small intestine extends from the pyloric sphincter to the large intestine. It is about 2 meters long in a living person and 6 to 7 meters long in a cadaver. It is called the **small** intestine not due to its length – the large intestine is much shorter – but due to its diameter, which is about 2.5 cm. It consists of three regions (measurements given here are based on the cadaver):

- The duodenum, the proximal portion, is about 25 cm long. It receives pancreatic juice, bile, and the contents of the stomach.
- The jejunum, the middle portion, is about 2.5 meters long. There is no conspicuous anatomical landmark that marks its end.
- The ileum, the distal portion, is about 3.6 meters long and connects to the cecum, the proximal portion of the large intestine. The ileocecal sphincter is located at this junction.

Numerous projections, called villi (singular: villus), which are about 0.5 to 1.5 mm long, extend from the surface of the mucosa into the lumen of the small intestine. The villi are largest in the duodenum and become progressively smaller in more distal regions of the small intestine. Villi greatly increase the surface area available for digestion and absorption. The core of a villus consists of lamina propria containing an arteriole, a venule, a blood capillary network, and a lacteal (a lymphatic capillary). The core is covered with two kinds of simple columnar epithelial cells: enterocytes and goblet cells. Enterocytes are absorptive cells. The apical plasma membrane of each cell has numerous microvilli that form a brush (striated) border about 1 μm high. By greatly increasing the surface area of the enterocytes, the microvilli increase the efficiency of absorption. In addition, some of the proteins in their plasma membranes are brush border enzymes, which are responsible for some of the final stages of chemical digestion. These enzymes are not released into the lumen – instead, the chyme must contact the brush border for the enzymes to function. Goblet cells are less abundant in the duodenum and become more frequent toward the ileum. They produce mucus that protects and lubricates the intestinal lining.

Between the bases of the villi are numerous openings of glands called intestinal glands (or crypts of Lieberkuhn). These glands, which extend as deep as the muscularis mucosae, secrete intestinal juice, which contains water and mucus and is slightly alkaline (pH = 7.4 to 7.8). The epithelium of the intestinal glands, which is continuous with that of the villi, contains stem cells, some absorptive cells, goblet cells, Paneth cells, and enteroendocrine cells. Stem cells dominate the lower region of the intestinal glands. They produce daughter cells that migrate upward and replace the enterocytes and goblet cells of the villi. After reaching the tip of the villi, they are sloughed off and digested. In this way, the entire epithelial lining of the intestine is replaced every 3 to 5 days. Paneth cells secrete lysozyme – an enzyme that digests the cell wall of some bacteria – and are capable of phagocytosis. These cells are located in the deepest parts of the intestinal glands.

The submucosa of the duodenum contains prominent duodenal (or Brunner's) glands that open into the intestinal glands. They secrete a distinctly alkaline mucus (pH = 8.1 to 9.3) that helps neutralize stomach acid in the chyme while shielding the mucosa from its corrosive effects.

In the region from the duodenum to the middle of the ileum – but primarily in the jejunum – the mucosa and submucosa form permanent circular folds, up to 1 cm high, called plicae circulares (or Kerckring's valves). They further increase the surface area and efficiency of absorption in this part of the digestive tract.

The lamina propria and submucosa of the small intestine contain solitary lymphoid nodules as well as aggregates, called Peyer's patches, of lymphoid nodules. Each patch, which is visible to the naked eye, consists of 10 to 200 nodules. There are about 30 Peyer's patches in the human body, most of them found in the ileum.

The muscularis consists of a relatively thick inner circular layer and a thinner outer longitudinal layer. It is responsible for two types of movement in the small intestine:

- Segmentation is the most common type, in which circular constrictions of the intestine cut into the contents, churning and mixing them with digestive juices and bringing nutrients into contact with the mucosa for absorption.
- Peristalsis propels undigested residue toward the large intestine.

Most of the duodenum is retroperitoneal – between the parietal peritoneum and posterior abdominal wall – and thus is covered with adventitia. However, the jejunum and ileum, lying within the peritoneum, are covered externally with serosa. This serosa is continuous with the common mesentery that suspends this portion of the small intestine from the posterior abdominal wall.

TOPIC 19.8. THE LARGE INTESTINE

The large intestine receives chyme from the small intestine and, by absorbing water and electrolytes, converts it into a solid or semisolid material called feces. It also eliminates feces from the body by the process of defecation. In the cadaver, the large intestine is about 1.5 meters long and 6.5 cm in diameter. It consists of four major regions: the cecum, the colon, the rectum, and the anal canal.

The cecum, a blind pouch in the lower right quadrant of the abdominal cavity, is the portion of the large intestine inferior to the ileocecal sphincter. The ileocecal sphincter, at the junction of the ileum and cecum, controls the movement of chyme from the small intestine into the large intestine. The vermiform ("worm-like") appendix, a blind tube with an average length of 8.3 cm, is attached to the lower end, or apex, of the cecum. The appendix is a modification of the cecum, rather than a distinct, separate organ. The human appendix and the end of the mammalian cecum have the same position, a similar structure and form, and a common developmental origin. Primates vary from having a large cecum without an appendix, to having a small cecum with an appendix.

The colon begins at the open end of the cecum at the ileocecal sphincter and can be divided into four segments. The ascending colon passes vertically along the right side of the abdominal cavity and makes a 90° turn when it reaches the inferior surface of the liver. The transverse colon begins at the superior end of the ascending colon, passes horizontally across the upper abdominal cavity, and turns 90° downward near the spleen. The descending colon begins at the distal end of the transverse colon and passes down the left side of the abdominal cavity to the level of the iliac crest. The sigmoid colon, roughly **S**-shaped, begins at the inferior end of the descending colon and passes medially to the midline.

The rectum passes about 12 cm downward from the sigmoid colon along the inferior half of the sacrum in a secondarily retroperitoneal position. It terminates at the anal canal. Just above its union with the anal canal, the rectum becomes dilated.

The anal canal is the last 2 to 3 cm of the digestive tract. Its opening to the exterior is called the anus, which is normally closed by an internal sphincter of smooth muscle tissue (involuntary) and an external sphincter of skeletal muscle tissue (voluntary). The anus opens during the process of defecation.

One of the distinctive characteristics of the wall of the large intestine concerns its mucosa. This layer, which lacks villi, has a simple columnar epithelium in all regions except the anal canal, where its epithelium is nonkeratinized stratified squamous. The latter provides protection against abrasion caused by the passage of feces. The simple columnar epithelium consists of mostly enterocytes and goblet cells. The enterocytes absorb water and the goblet cells secrete mucus that lubricates the passage of the colonic contents. Intestinal glands extend deeper into the mucosa and have a greater density of goblet cells than in the small intestine; mucus is their only significant secretion. The mucosa also contains solitary lymphoid nodules.

The muscularis of the large intestine, as in most of the digestive tract, consists of an outer longitudinal layer and an inner circular layer. However, the longitudinal layer between the vermiform appendix and the rectum is unusual in that its fibers do not form a continuous layer around the gut, but are collected in three longitudinal bands called taeniae coli, each about 12 mm wide. The bands are separated by portions of the wall with less or no longitudinal muscle at all; in the latter case, only the circular layer of smooth muscle is present between them. The taeniae coli are shorter than the other layers of the wall, causing it to form pouch-like regions called haustra. However, the rectum and anal canal lack haustra because there the longitudinal muscle forms a continuous sheet around the gut.

The ascending and descending colon are retroperitoneal, whereas the transverse and sigmoid colon are covered with serosa and are attached to the posterior abdominal wall by a mesentery called the mesocolon. The serosa often has epiploic appendages, which are small pouches of visceral peritoneum that are filled with fat and attached to the taeniae coli.

Despite its anatomical location, the vermiform appendix does **not** function as part of the human digestive system. In most vertebrates, the cecum is a large, complex gastrointestinal organ that is rich in gut-associated lymphoid tissue, or GALT (that is, MALT found in the gut), and is specialized for digestion of plants. In general, the relative size of the cecum is proportional to the amount of plant matter in a given organism's diet. The bacteria inhabiting the cecum secrete cellulase, an enzyme that digests cellulose. Even though humans eat plants, neither the human appendix nor the small cecum house cellulose-digesting bacteria, so humans cannot digest cellulose.

Even if the human appendix can be considered part of the lymphatic/immune system, the vast majority of GALT in humans is located outside the appendix in patches throughout the small and large intestines. The appendix can be removed without adverse complications and even congenital absence appears to have no discernable effect.

When the appendix is present, there is a 7 percent lifetime risk of acute appendicitis, which was usually fatal before the advent of modern surgical techniques.[90] Appendicitis occurs if the tiny opening to the appendix becomes blocked by kinking of the organ or with chyme, a foreign body, a tumor, etc., causing the appendix to swell. Death can result if the swollen appendix ruptures.

Therefore, the human appendix is a vestigial structure.[91] It has lost the digestive function it had in our distant ancestors, and at best plays a minor role in our modern lymphatic system. As Joseph McCabe wrote, "The vermiform appendage – in which some recent medical writers have vainly endeavoured to find a utility – is the shrunken remainder of a large and normal intestine of a remote ancestor. This interpretation of it would stand even if it were found to have a certain use in the human body. Vestigial organs are sometimes pressed into a secondary use when their original function has been lost."[92]

TOPIC 20. THE URINARY SYSTEM

The urinary system consists of two kidneys, two ureters, the urinary bladder, and the urethra. The kidneys maintain homeostasis by producing urine, which contains a variety of metabolic wastes. The rest of the urinary system functions to transport, store, and eliminate urine from the body. This ability to concentrate wastes and control water loss was crucial to the evolution of terrestrial animals such as humans. One of the most toxic metabolic wastes produced by the body and excreted mostly by the kidneys is a group of small nitrogen-containing compounds called nitrogenous wastes. These include ammonia, urea, uric acid, and creatinine. The kidneys also excrete toxins, drugs, hormones, salts, hydrogen ions, and water.

In addition, the kidneys regulate the fluid and electrolyte balance of the body and produce the enzyme renin and the hormone erythropoietin. Renin helps regulate blood pressure and erythropoietin stimulates the production of red blood cells. Other functions of the kidneys include synthesizing new glucose molecules (gluconeogenesis) during periods of fasting or starvation and participating in the synthesis of vitamin D.

The paired reddish kidneys are attached to the posterior abdominal wall just above waist level. Both the kidneys and ureters are retroperitoneal. The renal hilus is located near the center of each kidney's concave medial border and expands into a cavity within the kidney called the renal sinus. At the hilus, the ureter, blood vessels, lymphatic vessels, and nerves emerge from the kidney. The kidney is surrounded by three layers of tissue:

- The outer renal fascia, immediately deep to the parietal peritoneum, binds the kidney and associated organs to the posterior abdominal wall.
- The middle adipose capsule is a layer of fat that cushions the kidney and holds it in place.
- The inner renal capsule is a fibrous sac that is anchored at the hilus and encloses the rest of the kidney. It protects the kidney from injury and infection.

A frontal section through the kidney reveals two distinct regions: an outer layer about 1 cm thick called the renal cortex and an inner region called the renal medulla. The cortex is subdivided into an outer cortical zone and an inner juxtamedullary zone. The medulla contains 8 to 18 cone-shaped renal pyramids. They exhibit striations because they contain roughly parallel bundles of tiny urine-collecting ducts. The base of each pyramid is adjacent to the cortex and the pyramid's apex, called a renal papilla, points toward the hilus. Portions of the renal cortex that extend between renal pyramids are called renal columns. A renal lobe consists of a renal pyramid plus its overlying area of renal cortex and one-half of each adjacent renal column.

Each kidney contains about one million functional units called nephrons. Each nephron consists of two portions: a renal corpuscle and a renal tubule. The renal corpuscle is found in the renal cortex. The corpuscle, in turn, consists of two parts: the glomerulus and the glomerular capsule. The glomerulus is a network of capillaries, which receive blood from an afferent arteriole and are tributaries of an efferent arteriole. The glomerular (or Bowman's) capsule surrounds the glomerulus. It consists of two continuous layers of simple squamous epithelium, the outer parietal and inner visceral. Between the parietal and visceral layers lies the capsular (Bowman's) space. The modified epithelial cells of the visceral layer of the glomerular capsule are called podocytes and they completely encircle the glomerular capillaries. The podocytes and the endothelial cells of the glomerular capillaries form the filtration membrane, or endothelial-capsular membrane.

The capsular space in the renal corpuscle is continuous with the lumen of the renal tubule. The tubule, in turn, consists of three segments: the proximal convoluted tubule (PCT), the nephron loop (loop of Henle), and the distal convoluted tubule (DCT). Both convoluted tubules lie within the renal cortex, whereas the nephron loop extends into the renal medulla. The first portion of the nephron loop, the descending limb, passes from the cortex into the medulla. At its lower end, it turns 180° and forms the ascending limb that returns to the cortex. In each nephron, the final portion of the ascending limb of the nephron loop makes contact with the afferent arteriole serving that renal corpuscle, where the juxtaglomerular apparatus (JGA) is located. The JGA helps regulate blood pressure and the rate of blood filtration by the kidneys.

The efferent arterioles branch to form the peritubular (or cortical) capillaries and sometimes long loop-shaped capillaries called vasa recta (medullary capillaries). The peritubular capillaries surround tubular parts of the nephron mainly in the renal cortex, and the vasa recta surrounds tubular parts of the nephron in the renal medulla. Blood from the peritubular capillaries and vasa recta drains into the interlobular veins and eventually the renal vein.

There are two types of nephrons: cortical and juxtamedullary. The renal corpuscle of a cortical nephron (80 to 85 percent of all nephrons) lies in the outer portion of the cortex, its short nephron loop penetrating only into the outer region of the medulla. The short nephron loops are supplied with blood by peritubular capillaries. The renal corpuscle of a juxtamedullary nephron lies deep in the cortex, its long nephron loop penetrating deeply into the medulla. The long nephron loops are supplied with blood from peritubular capillaries and from the vasa recta.

A single collecting duct drains the DCTs of several nephrons. A collecting duct together with all the nephrons it drains is called a renal lobule. A single layer of epithelial cells forms the entire wall of the glomerular capsule, renal tubule, and collecting ducts.

Fluid in the nephron and collecting duct is given different names that reflect its changing composition: glomerular filtrate is the fluid in the capsular space, tubular fluid is the fluid in the renal tubule, and urine is the fluid in the collecting duct. Urine formation occurs continuously and involves three major processes: filtration, reabsorption, and secretion.

Filtration occurs in the renal corpuscle. Blood enters the glomerulus through the afferent arteriole. As blood flows through the glomerulus, it is filtered by the endothelial-capsular membrane. The capsular space collects the plasma-like filtrate, which consists of all blood components except formed elements and most proteins. The filtrate flows into the renal tubule. Blood leaves the glomerulus through the efferent arteriole.

The renal tubule and collecting duct are active in resorption. The normal rate of glomerular filtration is so high that, on average, the daily volume of fluid entering the PCTs is about 65 times the entire blood plasma volume. Thus, many useful solutes and about 99 percent of the water are reabsorbed from the tubular fluid and urine. The water and solutes return to the blood flowing through the peritubular capillaries and vasa recta. The PCT reabsorbs a greater variety of solutes, including glucose, amino acids, uric acid, urea, and various ions, than any other part of the nephron. The collecting duct reabsorbs water and concentrates the urine.

The renal tubule is also active in secretion. Some substances are transported from capillary blood into the tubular fluid. The substances secreted include hydrogen ions, potassium ions, nitrogenous wastes, and certain drugs such as aspirin. The DCT essentially completes the process of determining the chemical composition of the urine.

Collecting ducts within a renal pyramid merge to form papillary ducts, which open at the tips of the renal papilla to discharge urine into a minor calyx. Two or three minor calyces converge to form a major calyx, and two or three major calyces converge into a single large cavity called the renal pelvis. The calyces and part of the renal pelvis are located within the renal sinus.

Urine collected in the renal pelvis flows into a ureter that exits the kidney at the hilus and extends to the urinary bladder. The wall of each ureter consists of three layers: an inner mucosa, a middle muscularis, and an outer adventitia. The mucosa has a transitional epithelium continuous with that of the renal pelvis above and urinary bladder below. The muscularis consists of an inner longitudinal layer and an outer circular layer of smooth muscle tissue, with an extra, outer longitudinal layer of smooth muscle tissue in the distal third of the ureters. The adventitia has extensions that anchor the ureters in place. The ureters function to transport urine to the urinary bladder, which is accomplished through peristalsis by the muscularis, hydrostatic pressure of urine in the renal pelvis, and gravity.

The urinary bladder is a hollow muscular organ located in the pelvic cavity, posterior to the pubic symphysis and inferior to the peritoneum. In males, it is directly anterior to the rectum and usually has a larger capacity than in the female. In females, it is anterior to the vagina and inferior to the uterus. The shape of the urinary bladder ranges from flattened to spherical to pear-shaped, depending on the volume of urine it contains.

The wall of the urinary bladder consists of three layers: an inner mucosa, a middle muscularis, and an outer adventitia or serosa. The mucosa has a transitional epithelium, and in the relaxed bladder it has conspicuous wrinkles called rugae. As the bladder fills, it expands superiorly and the rugae flatten. The muscularis, called the detrusor muscle, consists of three layers of smooth muscle: an inner longitudinal, a middle circular, and an outer longitudinal. The bladder is covered by parietal peritoneum, or serosa, on its superior surface and by an adventitia elsewhere.

The floor of the urinary bladder has a small triangular area, called the trigone, marked by the two posterior ureteral openings and the anterior urethral opening, or internal urethral orifice. Around the opening to the urethra, the circular fibers of the detrusor muscle form an internal urethral sphincter. Since this sphincter is composed of smooth muscle, it is not under voluntary control. Where the urethra passes through the urogenital diaphragm, it is encircled by the external urethral sphincter. Since this sphincter is composed of skeletal muscle, it provides voluntary control over urination.

The urethra is a tube that conveys urine from the urinary bladder out of the body. Its opening to the outside is called the external urethral orifice. The wall of the urethra consists of two layers: an inner mucosa and an outer muscularis. In females, the urethra is a relatively vertical tube 3 to 4 cm long bound to the anterior wall of the vagina by fibrous connective tissue. Its external orifice lies between the vaginal orifice and clitoris. In males, the urethra is also a passageway for discharging semen from the body. It is about 15 to 20 cm long and its external orifice is located at the tip of the penis. The male urethra has three regions: the prostatic urethra, which passes through the prostate gland; the membranous urethra, which passes through the urogenital diaphragm; and the spongy (penile) urethra, which passes through the corpus spongiosum of the penis and ends at the external urethral orifice. The urethra in males has an **S**-shape. It passes downward from the bladder, turns anteriorly as it enters the root of the penis, and then turns about 90° downward again as it enters the external, pendant part of the penis.

The process of expelling urine is called micturition, urination, or voiding. Unlike urine formation, which is continuous, urination is episodic. When the volume of urine in the bladder reaches about 200 mL, baroreceptors (stretch receptors) in its wall send nerve impulses to sacral segments S2 to S3 of the spinal cord. This triggers a spinal reflex called the micturition reflex. Autonomic motor impulses cause contraction of the detrusor muscle and relaxation of the internal urethral sphincter, while somatic motor impulses to the external urethral sphincter are inhibited, causing it to relax. This is the predominant mechanism that voids the bladder in infants and young children. However, in childhood we acquire voluntary control over the external urethral sphincter, and emptying of the bladder is controlled predominantly by a micturition center in the pons. It is also possible to urinate when there is less than 200 mL of urine in the bladder by voluntarily contracting abdominal muscles that compress the bladder, thereby exciting the baroreceptors in its wall.

TOPIC 21. AN OVERVIEW OF THE REPRODUCTIVE SYSTEMS

Sexual reproduction is the process by which organisms produce offspring by the fusion of two gametes, or sex cells, usually from different parents. The gametes of humans are the mature egg, or ovum (plural: ova), and the sperm cell, or spermatozoon (plural: spermatozoa). Meiosis is the process that produces the gametes, and fertilization is the process in which two gametes fuse together. Sperm cells are produced by a type of meiosis called spermatogenesis and mature eggs by oogenesis. The essence of sexual reproduction is that each offspring has two parents and a combination of genes from both. As a result, the offspring are genetically different from their parents and usually from each other. This genetic diversity provides the individual variation required for natural selection. Thus, sexual reproduction is so important for the survival and evolution of a species that it is employed by the great majority of organisms.

The organs of the human reproductive systems may be grouped by function. The primary sex organs, or gonads, are those that produce the gametes. They also secrete sex hormones. The male gonads are the testes (singular: testis) and the female gonads are the ovaries. The secondary sex organs are organs other than the gonads that are essential to reproduction. In the male, these are the system of ducts, glands, and the penis, which provide for the storage, survival, and transport of sperm. In the female, they include the uterine tubes, uterus, and vagina, which provide a sheltered internal environment where fertilization and embryonic development of the fertilized egg, or zygote, can occur. The external genitalia occupy the perineum, a diamond-shaped region marked by the pubic symphysis, ischial tuberosities, and coccyx.

TOPIC 21.1. THE MALE REPRODUCTIVE SYSTEM

The male reproductive system consists of the:

- Testes or testicles (male gonads), which produce sperm and sex hormones.
- System of ducts that stores and transports sperm to the exterior.
- Accessory sex glands, which produce secretions that mix with sperm to form semen.
- Supporting structures, such as the scrotum and penis.

The scrotum in the human male (and in most other placentals and in marsupials) is a pendulous sac that encloses the two testes. It is an outpouching of the anterior abdominal wall consisting of loose skin and superficial fascia. A vertical median septum divides the scrotum into two compartments, each containing a testis. The septum consists of superficial fascia and smooth muscle tissue called the dartos. The dartos muscle is also found in the subcutaneous tissue of the scrotum, causing the skin of the scrotum to wrinkle. The location of the septum is externally marked by a seam called the perineal raphe, which also extends anteriorly along the ventral side of the penis and posteriorly as far as the margin of the anus.

The spermatic cord is a structure that extends from each testis up the back of the scrotum, then anterior to the pubis, and through the inguinal ring into the inguinal canal. The canal is a passage about 4 cm long through the anterior abdominal wall to the pelvic cavity. The cord contains blood and lymphatic vessels, autonomic nerves, and a spermatic duct called the ductus deferens (see below). The cremaster muscle consists of strips of the internal abdominal oblique muscle that enmesh the spermatic cord.

In the scrotum, the testes are kept at a temperature that is about 3° C lower than core body temperature. The cremaster muscle and dartos muscle help maintain this condition by adjusting the position of the testes. This lower temperature is essential for proper production and survival of sperm.

The testes are paired oval glands partially covered by a serous membrane called the tunica vaginalis. The white, fibrous capsule of the testis is called the tunica albuginea. Inward extensions (or septa) of the tunica albuginea divide each testis into 200 to 300 wedge-shaped compartments called lobules.

Each testicular lobule contains one to three tightly coiled seminiferous tubules where spermatogenesis occurs. Seminiferous tubules are lined with spermatogenic cells in various stages of development. Embedded between the spermatogenic cells are testicular sustentacular (Sertoli) cells, which extend from the basement membrane to the lumen. They support and protect developing spermatogenic cells. Tight junctions between sustentacular cells form the blood-testis barrier (BTB), which prevents the immune system from attacking the sperm cells. Between the seminiferous tubules are clusters of endocrinocytes called testicular interstitial cells (Leydig cells), which secrete testosterone.

The seminiferous tubules of each lobule converge to form a straight tubule that leads into a network called the rete testis, which lies in the posterior part of the testis. Sperm and the fluid produced by the sustentacular cells flow from the seminiferous tubules into the rete, where the sperm partially mature. However, sperm do not swim while they are in the male reproductive tract.

To reach the urethra after leaving each testis, the sperm flow through a series of spermatic ducts in the following sequence: efferent ductules, epididymis, ductus (vas) deferens, and ejaculatory duct. There are about 12 efferent ductules, which carry sperm from the rete testis to the epididymis. They have clusters of ciliated cells that help drive the sperm along.

The epididymis is about 7.5 cm long and adheres to the posterior side of the testis. The epididymis consists of a single coiled duct (ductus epididymis) embedded in fibrous connective tissue and, if straightened out, would be over 6 meters long. It reabsorbs about 90 percent of the fluid secreted by the sustentacular cells. The sperm mature in the epididymis. The process of spermatogenesis that produces a mature sperm cell takes about 74 days. Sperm mature at a rate of about 300 million per day and, once ejaculated, have a life expectancy of about 48 hours within the female reproductive tract. Sperm are stored in the epididymis and in the adjacent portion of the ductus deferens. Stored sperm remain fertile for 40 to 60 days, but if they are not ejaculated within this time, they disintegrate and are reabsorbed by the epididymis.

The ductus (vas) deferens is a muscular tube about 45 cm long and 2.5 mm in diameter. From the epididymis, it ascends within the spermatic cord, passes through the inguinal canal, pierces the anterior abdominal wall, and enters the pelvic cavity where it loops over the side and then down the posterior border of the urinary bladder. The ductus deferens from each testis widens into a terminal ampulla just before it joins the ejaculatory duct, located posterior to the urinary bladder. During ejaculation, the smooth muscle layers in the walls of the ductus epididymis and the ductus deferens perform peristalsis to propel sperm into the ejaculatory duct. The ejaculatory duct, the last of the spermatic ducts, is about 2 cm long. It passes through the prostate gland and empties into the urethra.

The male urethra is a passageway for semen and urine, although it cannot pass urine during ejaculation due to an autonomic reflex that simultaneously closes the internal urethral sphincter. Semen (seminal fluid) is the fluid that leaves the male body during ejaculation. It is a mixture of sperm cells (less than 10 percent by volume) and secretions from the accessory sex glands. The average number of sperm cells in semen is between 50 and 150 million per ml. A typical ejaculation discharges 2 to 5 ml of semen.

Accessory sex glands secrete most of the liquid portion of semen. As a result of these secretions, semen has a slightly alkaline pH of 7.2 to 7.6, which neutralizes the acidic environment in the female vagina. The accessory glands include two seminal vesicles, the single prostate, and two bulbourethral glands.

The pair of seminal vesicles or seminal glands is located posterior to the urinary bladder, one gland associated with each ductus deferens. They secrete an alkaline, viscous fluid that contains fructose, prostaglandins, and clotting proteins. Fructose, a sugar, provides a source of energy for sperm motility. This fluid enters the ejaculatory duct and makes up 60 percent of semen by volume.

The toroidal prostate gland is located inferior to the urinary bladder and surrounds the prostatic urethra, which passes through the gland. The prostate gland is present only in placental mammals, and among these mammals is absent in edentates, martens, otters, and badgers. The prostatic secretion is a milky, slightly acidic fluid that contains citric acid and several enzymes. This fluid enters the prostatic urethra and makes up about 25 percent of semen by volume.

The passage of the urethra through the prostate gland is another example of appallingly poor design. The prostate gland weighs about 20 g by age 20, and continues to grow at varying rates throughout life. Growth is minimal until about age 45, when the gland begins to measurably enlarge. By age 70, over 90 percent of men show some degree of benign prostatic hyperplasia (BPH). Consequently, the prostate gland compresses the urethra, obstructs the flow of urine, and may promote bladder and kidney infections. A much better arrangement would be to route the urethra **around** the prostate gland, which would not impede the function of either the gland or the urethra.

The pair of pea-sized bulbourethral (or Cowpers's) glands is located beneath the prostate gland, one gland on each side of the membranous urethra within the urogenital diaphragm. During sexual arousal, they secrete an alkaline, slippery fluid into the spongy urethra that lubricates the end of the penis. This fluid also neutralizes the acidity of residual urine in the urethra, which would be harmful to the sperm.

The penis is the intromittent organ that delivers semen into the vagina. About half of the penis is the externally visible portion, called the shaft, and the other half is an internal root, below the body surface. The typical dimensions of a flaccid (nonerect) shaft are 8 to 10 cm long and 3 cm in diameter, while those of an erect shaft are 13 to 18 cm long and 4 cm in diameter. Distally, the penis ends in a swollen head called the glans penis, the proximal margin of which is called the corona. The skin of the penis continues over the glans as the prepuce or foreskin, which is often removed by circumcision. Most of the penis consists of three cylindrical bodies called erectile tissues, so-named because when they fill with blood during sexual arousal, the penis becomes erect. A midventral erectile body called the corpus spongiosum encloses the spongy urethra. The expanded portion of its proximal end located in the root is called the bulb of the penis. Two dorsolateral erectile bodies called the corpora cavernosa (singular: corpus cavernosum) are separated from each other by a median septum. They diverge after they enter the root, each called a crus (plural: crura) of the penis.

In vertebrates, the testes begin development in the embryo near the kidney. The testes of all nonmammalian animals stay here, safe and deep within the abdominal cavity, and are called testicond testes. Although the testes of some mammals, such as elephants and manatees, are also testicond, in most mammals the testes migrate downward, or descend, as the embryo develops. Descended testes may remain within the body, such as in rhinos, whales, and seals, but in most cases they pass through the inguinal canal and into the scrotum, becoming scrotal testes. For example, the testes of humans and horses are descended and scrotal.

However, the location of the testes in a scrotum outside the abdominal wall exposes them to possible physical trauma. In addition, the descent of the testes creates a circuitous anatomical route for the spermatic duct. Although the duct passes ventral to the ureter, the testis migrates dorsal to the ureter, requiring the spermatic duct to loop around the ureter. This migration significantly lengthens the duct, rather than shortening it. A more efficient design would be a migration of the testis on the same side of the ureter as the spermatic duct, or even embryonic development of scrotal testes *in situ*. So why the present condition?

The common ancestor of the mammals may have inherited scrotal testes from mammal-like reptiles living around 260 million years ago. As these reptiles evolved the ability to generate body heat by their own metabolic activities (that is, endothermy), core body temperatures would have become too high for spermatogenesis. Descent of the testes into a scrotum may thus have evolved to cool them. However, because scrotal testes are more costly, the scrotum was lost in some mammal lineages as soon as an alternative solution to the problem of the temperature sensitivity of spermatogenesis could evolve. This explains why loss of the scrotum is more common among mammals than is loss of testis migration.   Phylogenetic analyses of class Mammalia support this hypothesis. [93, 94]

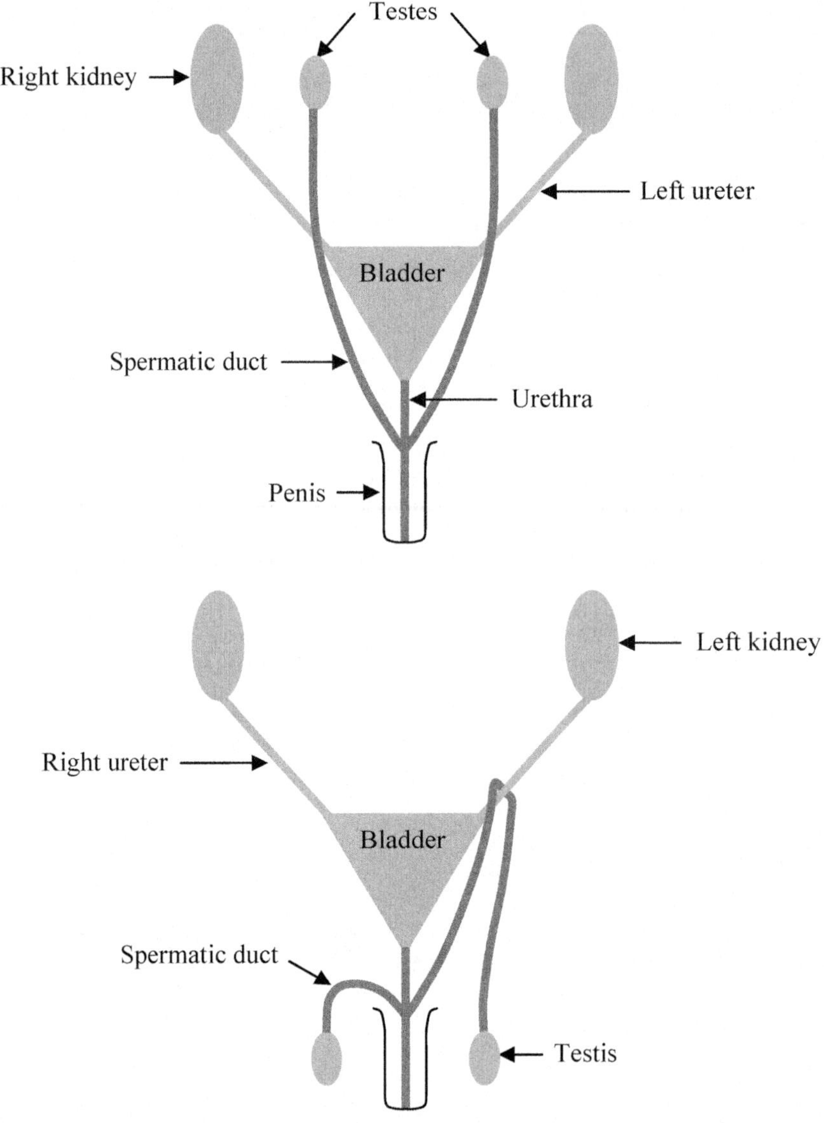

27. *The upper diagram shows a pair of testicond testes. The right side of the lower diagram shows a descended, scrotal testis. The left side of the lower diagram shows the configuration that would be expected if testis migration had been intelligently designed in a single, creative step.*

Mammals that have lost the scrotum have evolved alternative ways to cool the testes. For example, whales and seals keep their internal testes cool by shunting subcutaneous blood vessels back to the testes; in effect, they have "water-cooled balls". [95]

TOPIC 21.2. THE FEMALE REPRODUCTIVE SYSTEM

The female reproductive system consists of the:

- Ovaries (female gonads), which produce secondary oocytes (cells that develop into ova only after fertilization) and several hormones.
- Uterine (Fallopian) tubes or oviducts, which transport secondary oocytes and ova to the uterus.
- Uterus, in which embryonic and fetal development occur.
- Vagina, which leads from the exterior to the uterus.
- Vulva or pudenda, the external female genitalia.
- Accessory sex glands and erectile tissues.
- Mammary glands, which produce milk.

The two ovaries are located in the pelvic cavity, one on each side of the uterus and nestled in a depression of the posterior pelvic wall called the ovarian fossa. Each ovary has a hilus where blood vessels and nerves enter. An ovary is about 3 cm long, 1.5 cm wide and 1 cm thick. Its capsule of dense connective tissue, like that of the testis, is called the tunica albuginea, but without inward extensions (septa). Overlying the ovarian tunica albuginea is a simple cuboidal epithelium called the germinal epithelium, which was misnamed since it does not produce the germ cells – it is part of the parietal peritoneum.

The interior of the ovary is indistinctly divided into two regions: an outer ovarian cortex, containing ovarian follicles, and an inner ovarian medulla, containing blood vessels, lymphatic vessels, and nerves. Ovarian follicles consist of oocytes (immature eggs) in various stages of development and their surrounding epithelial cells. A mature or vesicular (Graafian) follicle is the last stage, consisting of a large fluid-filled follicle ready to rupture and expel a secondary oocyte in a process called ovulation. The corpus luteum is the remnant of the mature follicle after it has ruptured. It secretes several hormones until it degenerates into a mass of fibrous tissue called a corpus albicans.

Each ovary is held in position by a series of ligaments. The ovarian ligament attaches the ovary to the uterus. The suspensory ligament attaches the ovary to the pelvic wall and conducts major blood vessels and nerves to the hilus. The mesovarium, which is a short peritoneal fold, attaches the ovary to the broad ligament of the uterus. The broad ligament is a sheet of peritoneum that flanks the uterus on each side and encloses the oviduct in its superior margin, called the mesosalpinx.

Oogenesis begins in females before they are even born, unlike the process of spermatogenesis that begins in males only at puberty. Prior to birth, some diploid germ-line cells enlarge, forming diploid primary oocytes that begin meiosis I but do not complete it until after the female reaches puberty. Each of these is surrounded by a single layer of epithelial cells called follicular cells. At this stage, the primary oocyte plus the follicular cells is called a primordial follicle. Some of the primordial follicles develop into primary follicles, and then secondary follicles, even during childhood.

After puberty, meiosis resumes each month in several secondary follicles, although only one will become a mature follicle. The diploid primary oocyte completes meiosis I, producing two haploid daughter cells: a small, nonfunctional first polar body and a much larger secondary oocyte. The secondary oocyte begins meiosis II but does not complete it. The mature ovarian follicle consists of the secondary oocyte plus the surrounding follicular cells. Each month ovulation occurs, in which a mature follicle ruptures and releases its secondary oocyte. Normally this cell is swept into the oviduct. If fertilization does **not** occur, the secondary oocyte degenerates. If fertilization does occur, the secondary oocyte completes meiosis II, producing two haploid daughter cells: a small, nonfunctional second polar body and a much larger ovum. The nuclei of the sperm cell and the ovum then fuse, forming a diploid zygote that develops into the embryo. All polar bodies degenerate. Thus, oogenesis produces only one gamete from each germ-line cell, unlike spermatogenesis that produces four.

The two uterine (Fallopian) tubes or oviducts extend laterally from the uterus and transport secondary oocytes and zygotes from the ovaries to the uterus. Each uterine tube has three main portions: the infundibulum, which is the trumpet-shaped distal (ovarian) end; the ampulla, which is the long, middle portion; and the isthmus, which is the short, narrow, thick-walled proximal (uterine) end. The wall of the tube consists of three layers: an inner mucosa, a middle muscularis, and an outer serosa. The mucosa contains ciliated columnar epithelial cells, which help move the secondary oocyte along the tube, and a smaller number of secretory cells. The muscularis consists of smooth muscle tissue that performs peristalsis to also help move the secondary oocyte or the zygote into the uterus. It takes about 3 days for the secondary oocyte to travel the length of the uterine tube, but it lives only 24 hours. If not fertilized, it will die before it arrives in the uterus. The serosa is a serous membrane continuous with the mesosalpinx, part of the broad ligament of the uterus.

The uterus (womb) is shaped like an inverted pear and is located between the urinary bladder and the rectum. It can be divided into four regions: the fundus, the dome-shaped region above the uterine tubes; the body (corpus), the major, tapering, central portion; the cervix, the lower narrow portion that opens into the vagina; and the isthmus, a constricted region connecting the body and the cervix. The uterus measures about 7 cm from cervix to fundus, 4 cm wide at its broadest point, and 2.5 cm thick, but it is somewhat larger in women who have been pregnant. The lumen of the uterus can likewise be divided into four parts: the uterine cavity, the interior space within the body; the cervical canal, the interior space within the cervix; the internal os, the junction of the uterine cavity with the cervical canal; and the external os, the opening of the cervical canal into the vagina. The wall of the uterus consists of three layers:

- The endometrium is the highly-vascular mucosa, which is divided into two layers:

an inner stratum functionalis (functional layer) that is shed during menstruation; and an outer stratum basalis (basal layer), a permanent layer that produces a new stratum functionalis after each menstruation.

- The myometrium constitutes most of the uterine wall and consists of three layers of smooth muscle tissue. It is thickest in the fundus and thinnest in the cervix. Its coordinated contractions during childbirth help expel the fetus.
- The perimetrium is the serosa, which is part of the visceral peritoneum.

The uterus is supported mainly by the muscular pelvic diaphragm and urogenital diaphragm but also by ligaments formed by folds of peritoneum. The broad ligament, mentioned earlier, has two parts: the mesosalpinx, the superior margin enclosing each oviduct, and the mesometrium, the portion flanking each side of the uterus. A pair of uterosacral ligaments attaches the posterior side of the uterus to the sacrum. The cardinal (lateral cervical) ligaments connect the cervix and the vagina to the pelvic wall. Two round ligaments extend from the anterior surface of the uterus, pass through the inguinal canals, and terminate in the labia majora of the external genitalia.

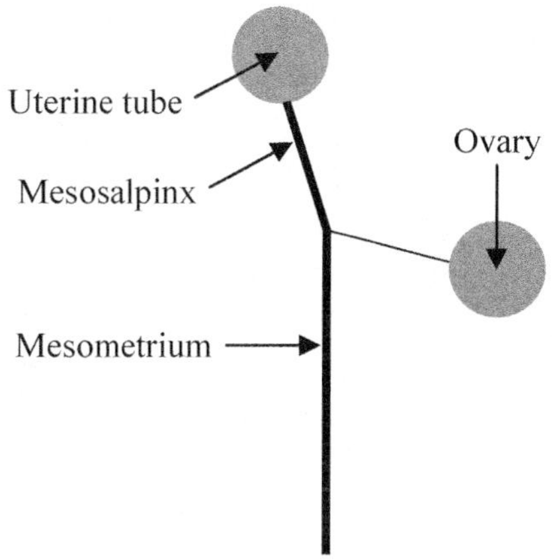

Uterine tube
Ovary
Mesosalpinx
Mesometrium

28. Schematic lateral view of ovary and uterine tube, showing relationship with broad ligament (thick line) and mesovarium (thin line).

The functions of the uterus include: menstruation; provision of a conduit for sperm to enter the oviducts; implantation, the attachment of a developing embryo to the endometrium; development of the embryo and fetus; and labor.

The uterus is another example of structural design **not** intelligently planned. As with the lower abdominal wall (see Topic 14), the erect posture of humans consequently increases gravitational stress on the pelvic diaphragm due to the weight of the viscera above. Pregnancy, childbirth, obesity, and/or the normal ageing process can weaken the pelvic diaphragm and ligaments and this can result in a prolapse (pelvic floor hernia). A prolapse occurs when the uterus or urinary bladder protrudes into the vaginal canal. Some doctors estimate that half of all women have some degree of uterine or bladder prolapse in the years following childbirth, although only 10 to 20 percent of women with this condition seek medical evaluation for symptoms.

In addition, it has been estimated that 80 percent of human conceptions end in spontaneous abortion or miscarriage.[96] The majority pass unnoticed because they occur within the first three weeks of pregnancy. Many of these spontaneous abortions appear to be due to an abnormal number of chromosomes, which may result in abnormal embryonic development.[97] Other miscarriages may be due to the poor health of the mother. Spontaneous abortion makes sense from an adaptive point of view.[98] If the fetus is in poor condition, it will be less likely to survive. If the mother is in poor condition, she may be unable to provide the resources for her offspring to survive to birth or beyond. When the body terminates a pregnancy early, time and energy that can be invested in future offspring are conserved. However, the wastefulness of human reproduction does not make sense if the process is attributed to a so-called Intelligent Designer.

The vagina (birth canal) is a tube about 8 to 10 cm long, which is located between the urinary bladder and the rectum and passes through the pelvic diaphragm. It allows for the discharge of menstrual fluid, receives the penis and semen during sexual intercourse, and provides a conduit for childbirth, also called delivery or parturition. The vagina extends slightly superiorly to the cervix, forming a recess called the fornix. The opening of the vagina to the exterior, called the vaginal orifice, is partially or completely covered by a thin fold of vascularized mucous membrane called the hymen. It has one or more openings to permit the passage of menstrual fluid, but it is ruptured during the first intercourse if not already ruptured by tampons, medical examinations, or strenuous exercise. The thin but very distensible wall of the vagina consists of three layers:

- The inner mucosa of the lower end of the vagina has folds called rugae, which stimulate the penis and help induce ejaculation. The vaginal epithelium is simple cuboidal in childhood, but is transformed into nonkeratinized stratified squamous epithelium during puberty. The epithelial cells are rich in glycogen, which bacteria ferment to lactic acid, creating a low pH (about 3.5–4.0) in the lumen. This acidity retards microbial growth but is neutralized by the semen so it does not harm the sperm.
- The middle muscularis can stretch considerably to receive the penis and permit passage of the fetus. It consists of an inner circular layer and an outer longitudinal layer of smooth muscle.
- The outer adventitia anchors the vagina to neighboring organs.

The vulva or pudenda consists of the external genitalia of the female. It includes the mons pubis, labia majora and minora, and clitoris. The mons pubis is an elevation of adipose tissue covered by skin and pubic hair that cushions the pubic symphysis. It is located anterior to the vaginal and urethral openings. The labia majora (singular: labium majus) consist of two longitudinal folds of skin that extend inferiorly and posteriorly from the mons pubis. They are covered by pubic hair and contain adipose tissue, sebaceous glands, and sudoriferous glands. The labia minora (singular: labium minus) consist of two smaller folds of skin located medial to the labia majora. They have some sudoriferous glands and many sebaceous glands. The region between the labia minora, called the vestibule, contains the vaginal orifice. The clitoris is a small cylindrical mass of erectile tissue and nerves located at the anterior junction of the labia minora. The external urethral orifice is located in the vestibule anterior to the vaginal orifice and posterior to the clitoris.

Beneath the skin of the perineum are several accessory glands and erectile tissues. Two mucus-secreting greater vestibular glands open into the vestibule near the vaginal orifice and provide most of the lubrication for intercourse. Two mucus-secreting paraurethral glands open into the vestibule near the external urethral orifice. Just deep to the labia majora on either side of the vaginal orifice are located two elongated masses of erectile tissue that form the vestibular bulb, or bulb of the vestibule. They fill with blood during sexual excitement and cause the vagina to tighten somewhat around the penis, enhancing sexual stimulation.

Certain structures have the same embryological origin in both sexes but perform different functions; therefore they are said to be homologous:

- The ovaries and testes.
- The ovum and spermatozoon.
- The labia majora and scrotum.
- The labia minora and spongy urethra.
- The vestibule and membranous urethra.
- The bulb of the vestibule and the corpus spongiosum and bulb of the penis.
- The clitoris and glans of the penis.
- The greater vestibular glands and the bulbourethral glands.
- The paraurethral glands and the prostate gland.

The mammary glands are modified sudoriferous glands that secrete milk. The function of the mammary glands is lactation. Lactation includes milk synthesis and secretion, which is stimulated primarily by the hormone prolactin, and milk ejection, which is stimulated by the hormone oxytocin. The mammary glands occupy the breasts, which are anterior to the pectoralis major and serratus anterior muscles and are attached to them by a layer of connective tissue. Each breast has two principle regions: the conical to pendulous body, with the nipple at its apex, and an extension toward the armpit called the axillary tail. The nipple is surrounded by a circular, colored zone called the areola, which has sparse hairs and areolar glands that are visible as small bumps on the surface. Areolar glands are developmentally intermediate between sudoriferous glands and mammary glands, producing a secretion during nursing that minimizes chapping and cracking of the areola. Internally, the nonlactating breast consists mostly of adipose and collagenous tissue with very little mammary gland. However, it does have a system of ducts that branches through its connective tissue and converges on the nipple. During pregnancy, the mammary gland develops, forming 15 to 20 lobes arranged radially around the nipple. Each lobe is drained by a lactiferous duct, which widens to form a lactiferous sinus opening onto the nipple.

The female reproductive cycle includes the uterine and ovarian cycles, the hormonal changes that regulate them, and cyclical changes in the breasts and cervix. The ovarian cycle is the series of events that occurs in the ovaries, and the uterine (menstrual) cycle is the series of changes that occurs in the endometrium. The duration of the reproductive cycle in a nonpregnant female is typically 28 days but can vary from 24 to 35 days. The female reproductive cycle can be divided into four phases:

- The menstrual phase (menstruation or menses) lasts about the first five days. In the ovaries, about 20 secondary follicles begin to develop. In the uterus, the entire stratum functionalis is shed, causing the discharge of menstrual flow through the vagina to the exterior.
- The preovulatory phase generally lasts from days 6 to 13 in a 28-day cycle. In the ovaries, one of the developing secondary follicles becomes a mature follicle. In the uterus, cells of the stratum basalis divide mitotically to produce a new stratum functionalis.
- Ovulation usually occurs on day 14 in a 28-day cycle. It is triggered by a surge in secretion of luteinizing hormone (LH). The mature follicle ruptures and the secondary oocyte is released into the pelvic cavity.
- The postovulatory phase generally lasts from days 15 to 28 in a 28-day cycle. In the ovaries, remnants of the mature follicle develop into a corpus luteum that secretes progesterone, estrogen, relaxin, and inhibin.

If fertilization does **not** occur, the corpus luteum degenerates into a corpus albicans, causing a decline in the secretion of progesterone and estrogen, which causes menstruation. If fertilization does occur, the embryo secretes human chorionic gonadotropin (hCG), which maintains the corpus luteum and its secretion of progesterone and estrogen, which causes the endometrium to prepare for the possible implantation of a developing embryo.

Other terms are sometimes used for phases of the female reproductive cycle. The menstrual and preovulatory phases together are called the follicular phase of the ovarian cycle. The preovulatory phase is also called the proliferative phase of the uterine cycle. The postovulatory phase is also called the luteal phase of the ovarian cycle and the secretory phase of the uterine cycle.

TOPIC 22. DEVELOPMENTAL ANATOMY: THE PRENATAL PERIOD

Developmental anatomy is the study of the sequence of events from the fertilization of a secondary oocyte (conception) to the formation of an adult organism. The term conceptus refers to any prenatal product of fertilization, ranging from a zygote to the full-term fetus, including its extraembryonic membranes, placenta, and umbilical cord.

Human prenatal development can be divided into two phases: embryogenesis and fetal growth (or phenogenesis). Embryogenesis begins at the moment of fertilization and lasts to the end of the eighth week of development. It, in turn, can be divided into four stages: fertilization, cleavage, gastrulation, and organogenesis. Cleavage refers to the rapid mitotic cell divisions of the zygote into progressively smaller cells, called blastomeres. This process begins about 30 hours after fertilization. Gastrulation is a series of extensive cell rearrangements that produces the three primary germ layers of the embryo. Organogenesis is the process in which the cells of the three germ layers interact with one another and rearrange themselves to produce the bodily organs. It begins with the process of neurulation. Fetal growth lasts from the end of embryogenesis until 40 weeks, when the fetus is born. This period can be compared to metamorphosis (for example, larval stages) in other animals in which embryogenesis culminates in hatching or birth.

During fertilization, the secondary oocyte and the sperm fuse. Within 11 hours following fertilization, the secondary oocyte completes meiosis II. The subsequent fusion of the ovum nucleus and sperm nucleus marks the creation of the zygote and the end of fertilization. Since a secondary oocyte must be fertilized within 12 to 24 hours after ovulation in order to survive, and it takes about 72 hours to reach the uterus, fertilization must occur in the oviducts.

During cleavage, no period of cell growth occurs between divisions. Therefore, the conceptus at this stage is no larger than the original zygote. Cleavage occurs in the oviduct as the conceptus migrates toward the uterus. The conceptus is called a morula (from the Latin for mulberry, descriptive of its appearance) from the time it has 16 blastomeres until the next stage, which is the blastula in nonmammalian animals. In mammals, rearrangement of cells of the morula produces the blastocyst, which is similar in structure to the blastula that develops from the morula of nonmammalian animals. A blastula consists of a spherical layer of around 128 cells surrounding a central fluid-filled cavity called the blastocoel.

However, the cells of a blastula and a blastocyst have different fates. The cells of a blastula give rise to all later structures of the adult organism, but a blastocyst consists of two primary cell lines: the **trophoblast**, which is the outer layer of squamous cells surrounding the blastocoel, and the **embryoblast** (also called the inner cell mass), which is an inner mass of cells located on one side of the blastocoel. The side of the blastocyst where the embryoblast is situated is called the **embryonic pole** and the diametrically opposite side is the **abembryonic pole**

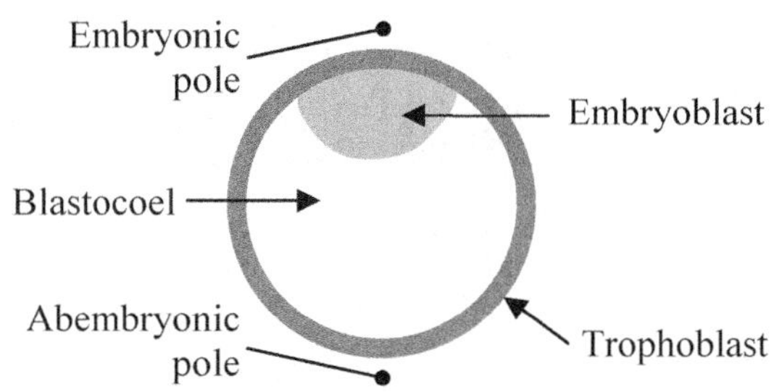

29. Schematic drawing of a preimplantation blastocyst (day 4 after fertilization).

(see Figure 29). Only the embryoblast contains the cells that give rise to all later structures of the adult organism, while the trophoblast interacts with the endometrium to eventually form the embryonic portion of the placenta, which is discarded at birth. Thus, the trophoblast is an adaptation for intrauterine development in placental mammals.

A blastocyst must embed itself in the wall of the uterus in order to survive, a process called implantation. Not all blastocysts are successful – one study of healthy women trying to conceive found 22 percent of pregnancies (identified by sensitive hormone measures) were lost prior to becoming detectable clinically. Even after implantation, there is a substantial rate of spontaneous abortion, still not known precisely but estimated at 25 to 40 percent.[99]

On or around the 6th day of development in humans, the trophoblast at the embryonic pole of the blastocyst loosely attaches to the endometrium. By the 8th day the cells of the trophoblast in contact with the endometrium give rise to a population of cells in which mitosis occurs in the absence of cytokinesis. These multinucleate cells form a layer called the **syncytiotrophoblast**, while the layer containing the initial type of trophoblast cell is now called the **cytotrophoblast**. The syncytiotrophoblast invades the endometrium by digesting endometrial cells, which provides nourishment for the embryoblast and enables the blastocyst to penetrate the uterine lining. By the 11th to 12th day the blastocyst becomes completely buried in the wall of the uterus. At this point, the entire trophoblast has differentiated into the internal layer of cytotrophoblast and the external layer of syncytiotrophoblast. After implantation, the endometrium is known as the decidua. Implantation is completed by the time the next menstrual period would have commenced had fertilization not taken place. The trophoblast prevents menstruation by secreting human chorionic gonadotropin (hCG), which prevents the corpus luteum from degenerating into the corpus albicans. As a result, the corpus luteum continues to secrete estrogen and progesterone, which maintains the decidua.

Coincident with implantation, the cells of the embryoblast are differentiating, producing two layers of cells called the epiblast and hypoblast. These terms originated from the early study of cleavage in birds, fishes, and reptiles, in which the large concentrations of yolk prevent cleavage in all but a small portion of the zygote cytoplasm. This restricted cleavage produces a tiny, flattened embryo composed of two cell layers, an upper epiblast and lower hypoblast, lying on an enormous mass of yolk. In mammals, the **hypoblast** (also called the primitive endoderm) is the surface layer of the embryoblast next to the blastocoel. Some of the hypoblastic cells migrate to line the blastocoel, where they give rise to an extraembryonic membrane called the **yolk sac membrane** (exocoelomic membrane). This process converts the blastocoel into the **yolk sac** (exocoelomic cavity). The remaining embryoblast, between the hypoblast and the cytotrophoblast, is now the **epiblast** (also called the primitive ectoderm). A small cavity arises within the epiblast and eventually enlarges to form the **amniotic cavity**. The layer of epiblastic cells between the amniotic cavity and the cytotrophoblast forms another extraembryonic membrane called the **amnion**, which will secrete amniotic fluid into the cavity. The remaining epiblast that does not contribute to forming the amnion is now called the **embryonic epiblast**. It contains all the cells that will generate the actual embryo.

As a result of these developments, a **bilaminar** (two-layered) **embryonic disc**, composed of hypoblast and embryonic epiblast, now separates the yolk sac and the amniotic cavity. However, despite its name, the yolk sac contains no yolk. In humans, the yolk sac is vestigial and most likely has a nutritive role only in early stages of development. In the second month of development, it lies in the chorionic cavity.

On the 11th to 12th day of development in humans a new population of cells, derived from yolk sac cells, fills the space between the cytotrophoblast and the amnion and yolk sac, forming the **extraembryonic mesoderm** (see Figure 30). By the 13th day a fluid-filled cavity called the **chorionic cavity** (extraembryonic coelom) arises within the extraembryonic mesoderm. This space surrounds the yolk sac membrane and amnion, except where the bilaminar embryonic disc is connected to the cytotrophoblast by the connecting stalk (see Figure 31). With development of blood vessels, the stalk will become the umbilical cord, which will connect the fetus to the placenta. The extraembryonic mesoderm lining the cytotrophoblast, the cytotrophoblast, and the syncytiotrophoblast all combine to form a third extraembryonic membrane called the **chorion**. By the end of the second month of pregnancy, the chorion begins producing estrogen and progesterone, taking over the role of the corpus luteum and making hCG unnecessary. The ovaries then become inactive for the rest of the pregnancy. The chorion also produces secretions that suppress the immune system of the mother so that she will not reject the embryo as she would an organ graft.

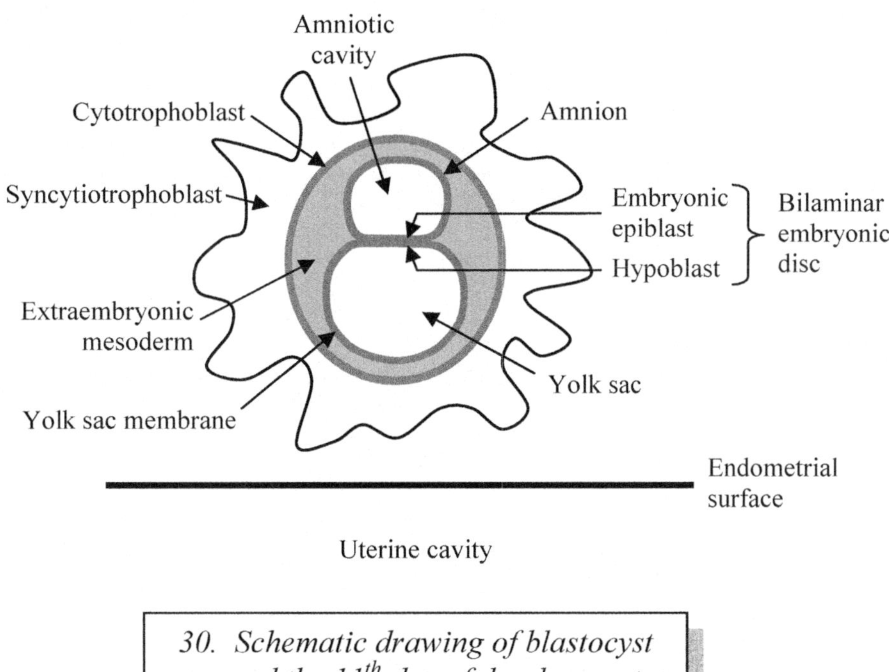

30. Schematic drawing of blastocyst around the 11th day of development.

Prior to gastrulation, the majority of cells (about 75 percent) in the preimplantation blastocyst comprise the trophoblast and hypoblast, which will never become part of the embryo. In fact, the human embryo on day 5 after fertilization contains 200 to 250 cells, only 30 to 34 of which are embryoblastic cells.[100]

During gastrulation, the embryonic epiblast becomes organized into the three primary germ layers of the embryo – ectoderm, endoderm, and mesoderm. The movements of cells during reptilian and avian gastrulation evolved as an adaptation to yolky eggs but, surprisingly, these movements are retained even in the absence of large amounts of yolk in the mammalian embryo. Thus, mammalian gastrulation movements are another example of an adaptive compromise.

The major structure characteristic of gastrulation in birds, reptiles, and mammals is the **primitive streak**, a faint groove on the dorsal (epiblastic) surface of the embryonic disc that elongates caudally (toward the tail end of the embryo). From this point on, one can identify the cephalic (head)/caudal (tail) ends, the dorsal/ventral sides, and the right/left sides of the embryo. At the cephalic end of the primitive streak, a small group of epiblastic cells forms a rounded structure called the **primitive node** (or Hensen's node). In humans, the primitive streak appears on the 14th to 16th day of development.

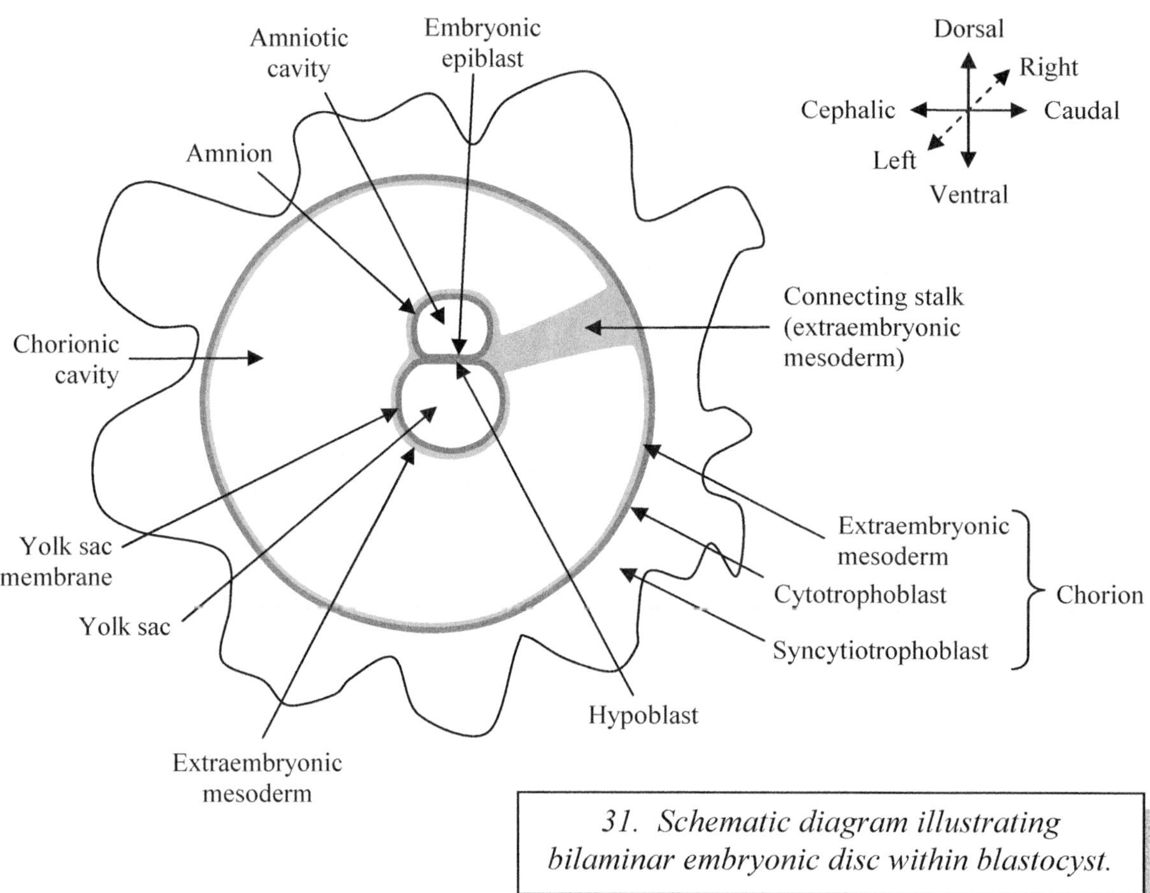

31. Schematic diagram illustrating bilaminar embryonic disc within blastocyst.

Some of the epiblastic cells migrate to the primitive streak, detach from the epiblast, and move into the middle of the embryonic disc. This inward movement is known as invagination. Once the cells have invaginated, some of them displace the hypoblast, forming the **embryonic endoderm**, and others come to lie between the epiblast and newly created endoderm, forming the **embryonic mesoderm**. Cells remaining in the epiblast become the **embryonic ectoderm** of the embryo. As a result of these movements, the bilaminar embryonic disc is converted into a **trilaminar** (three-layered) **embryonic disc**. In general, the fates of the three primary germ layers of the embryo are:

- The embryonic ectoderm gives rise to: the central nervous system and peripheral nervous system; outer surface or skin of the organism; cornea and lens of the eye; epithelium that lines the mouth and nasal cavities and the anal canal; epithelium of the pineal gland, pituitary gland, and adrenal medulla; and cells of the neural crest (which gives rise to various facial structures, melanocytes, and dorsal root ganglia).

- The embryonic mesoderm gives rise to: skeletal, smooth, and cardiac muscle; structures of the urogenital system (kidneys, ureters, gonads, and reproductive ducts); bone marrow and blood; bone, cartilage, and adipose and other connective tissues; and the lining of the body cavity.
- The embryonic endoderm gives rise to: the epithelium of the entire digestive tract (excluding the mouth and anal canal); epithelium of the respiratory tract; structures associated with the digestive tract (liver and pancreas); thyroid, parathyroid, and thymus glands; epithelium of the reproductive ducts and glands; and the epithelium of the urethra and bladder.

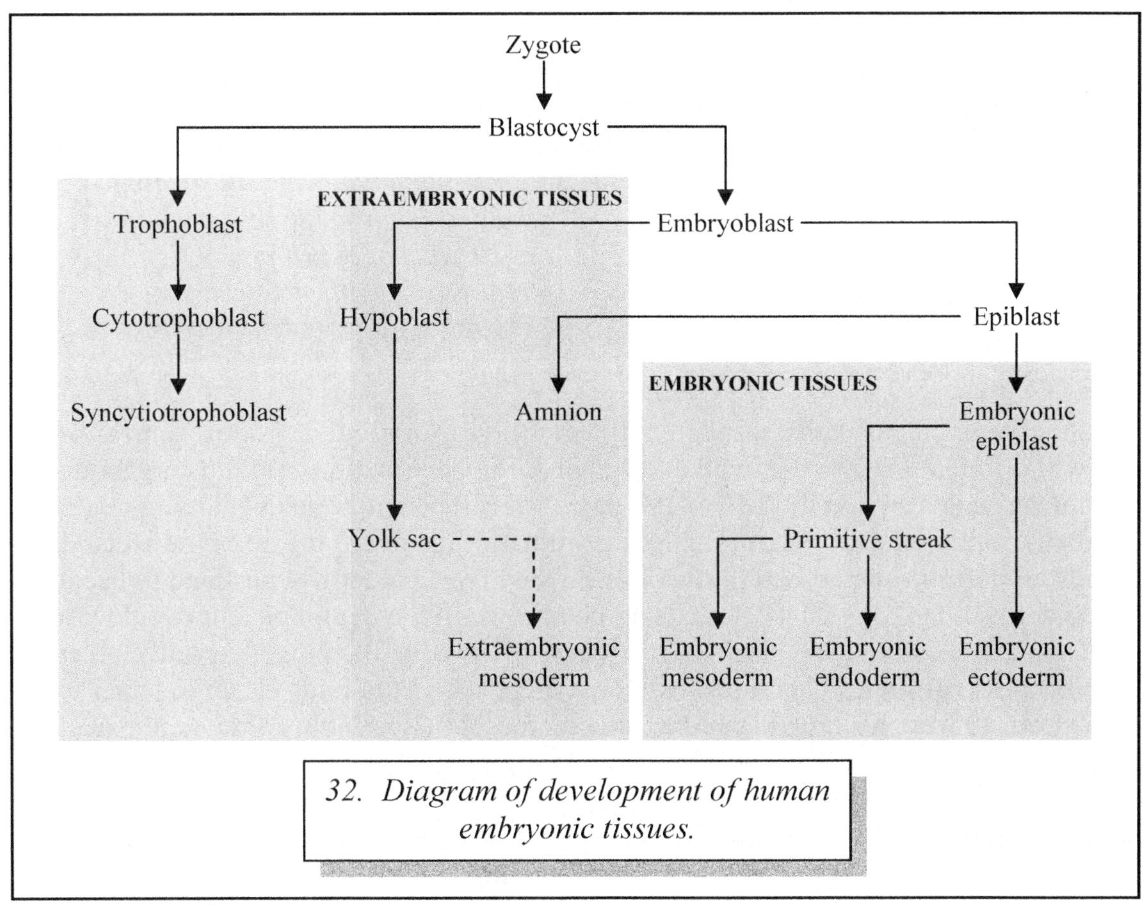

32. *Diagram of development of human embryonic tissues.*

On or about the 16th day of development in humans, cells from the primitive node migrate cranially (toward the head) just beneath the epiblast, forming a hollow tube of cells in the midline of the embryonic disc. By the 22nd to 24th day, this tube becomes a solid, pliant, longitudinal rod of connective tissue called the **notochord** (see Figure 33).

The **allantois** is the last extraembryonic membrane to form. It appears around the 16th day of development in humans as an outpocketing of the wall of the yolk sac near the caudal end of the embryonic disc and extends into the connecting stalk. In some vertebrates, the allantois is used to sequester excretion products of the renal system. Due to the role of the placenta in these functions, the human allantois remains rudimentary, although it does contribute to formation of the umbilical cord and development of the urinary bladder.

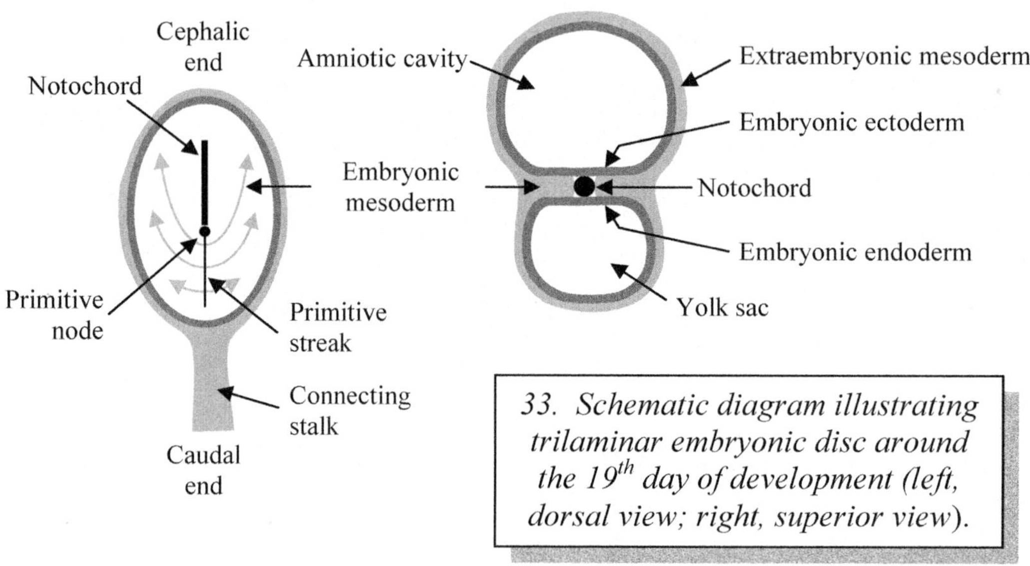

33. *Schematic diagram illustrating trilaminar embryonic disc around the 19th day of development (left, dorsal view; right, superior view).*

Organogenesis begins with neurulation, the process of forming the hollow **neural tube** that will differentiate into the brain and spinal cord. Neurulation, and thus organogenesis, in humans begins around the 18th to 19th day of development. The developing notochord within the embryonic mesoderm induces (stimulates) the overlying embryonic ectoderm to thicken, forming the **neural plate**. The surrounding ectoderm is destined to become the epidermis. On days 20 to 21, cells at the midline of the neural plate move downward and form a depression called the **neural groove**. The cells flanking the groove on each side form **neural folds**. Continued cell movement brings the folds closer together until they meet and fuse, forming the neural tube by the 27th day. In this process, the neural tube comes to lie below the (dorsal) surface (see Figure 34). The ectoderm overlying it will form the outer layer of skin. The cephalic portion of the neural tube will grow and differentiate into the brain, while the remainder of the tube will develop into the spinal cord. Eventually, the vertebral column will develop around the notochord, which will break up into sections that will form the weight-bearing intervertebral discs.

The mesoderm still adjacent to the neural tube forms paired blocks on either side of the tube, which are called **somites**. The first pair of somites appears on approximately the 20th day of development, at the cephalic end of the neural tube. More pairs appear in the caudal direction, up until about day 30. Mesodermal cells from the somites give rise to most of the skeleton and skeletal muscle.

A process called embryonic folding converts the embryo from a flat, two-dimensional trilaminar embryonic disc into a curved three-dimensional cylinder. Lateral folding forms the ventral body wall of the embryo. This occurs as the lateral margins of the trilaminar embryonic disc bend ventrally due to growth of the somites. These folds move toward the midline and incorporate the dorsal part of the yolk sac into the embryo as the **primitive gut** (archenteron). At the same time, the folding also forms the **intraembryonic body cavity**, which later develops into the ventral body cavity (see Figure 35). Cephalocaudal folding produces the **head fold** and the **tail fold**. Consequently, the cylindrical embryo becomes **C**-shaped. The head fold and tail folds bring the developing heart and mouth and the developing anus into their eventual adult positions, respectively. They also bring about expansion of the amniotic cavity to totally enclose the embryo.

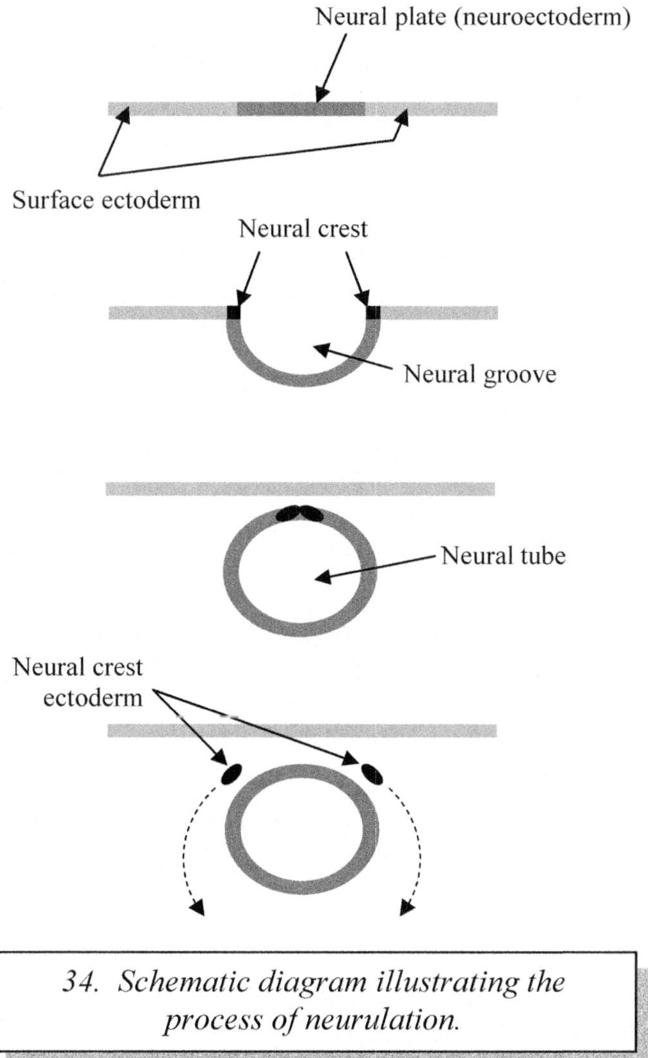

34. Schematic diagram illustrating the process of neurulation.

The primitive gut differentiates into an anterior **foregut**, an intermediate **midgut**, and a posterior **hindgut**. The foregut gives rise to: the pharynx; lower respiratory system;

esophagus; stomach; proximal half of the duodenum; liver and pancreas; and gallbladder and bile ducts. The midgut gives rise to: the distal half of the duodenum; jejunum and ileum; cecum and vermiform appendix; ascending colon; and most of the transverse colon. The hindgut gives rise to: the left part of the transverse colon; descending colon; sigmoid colon; rectum; superior part of anal canal; epithelium of urinary bladder; and most of the urethra.

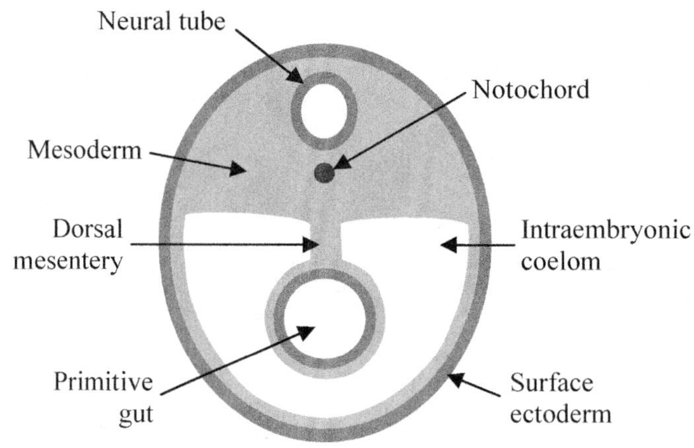

35. *Schematic diagram of transverse section just inferior to midgut of embryo after lateral folding.*

During organogenesis, the conceptus gradually switches from trophoblastic nutrition, in which it is nourished by digestion of endometrial cells by the syncytiotrophoblast, to placental nutrition, in which it is nourished by diffusion of nutrients from the mother's bloodstream through the placenta. The chorion forms the embryonic portion of the placenta, a vascular organ that develops on the uterine wall. Formation of the placenta extends from the 11th day through the 12th week of development but occurs primarily during organogenesis.

When all organ systems are present, embryogenesis is considered complete – the embryo is now called a fetus. During fetal growth, organs grow and mature at a cellular level to the point of being capable of supporting life outside the maternal body.

The presence of four anatomical features in the human embryo as well as in all chordates, a group of animals that consists of all vertebrate and a few invertebrate species, indicates that they are all descended from a common ancestor.

First, all chordates have a notochord during early development, which induces formation of the dorsal hollow nerve cord. The notochord or remnants thereof, persist into the adult stages of a group of chordates known as cephalochordates (amphioxus) and many vertebrates. It is semi-flexible and can bend laterally. It supports the axis of the body, extending into a muscular, post-anal tail. Lengthwise muscular contraction on either side of the notochord causes the tail to bend and propels an aquatic chordate forward. After it induces certain mesodermal cells to form vertebrae, most of the notochord degenerates in the later embryonic stages of most vertebrates. However, in between the vertebrae, the notochordal cells form the tissue of the intervertebral discs, called the nuclei pulposi. Also, the muscular, post-anal tail disappears during embryonic growth in apes and humans.

Also, the embryos of all chordates have a dorsal hollow nerve cord (neural tube), located just above and parallel to the notochord and under the body surface. The neural tube persists throughout adult life in almost all chordates. The brain and spinal cord develop from the neural tube itself, while the cavity of the neural tube is modified to form the ventricles of the brain and the central canal of the spinal cord.

Another characteristic feature in the embryos of all chordates is a series of pouches (lined by endoderm) that develops in the lateral walls of the pharynx. Over these pouches, corresponding indentations of the outer covering of the body occur, forming the branchial or pharyngeal grooves or clefts (lined by ectoderm). The pharyngeal pouches push through the mesoderm until they meet the ectoderm, with which they fuse to form thin membranes. The pharyngeal grooves separate a series of rounded bars, called the branchial or pharyngeal (visceral) arches, in which thickening of the mesoderm takes place. Each pharyngeal arch contains a cartilaginous bar-like structure, an artery, and a cranial nerve. In humans, six arches make their appearance, but of these only the first four are visible externally.

In fishes and amphibians, the membranes between the pharyngeal arches disappear and thus form a series of perforations, or gill slits, while the pharyngeal arches become gill arches. In birds and mammals, the pouches form in the characteristic way but the membranes do not disappear. The German anatomist Martin Heinrich Rathke (1793-1860) first described these embryonic precursors of gill slits and gill arches in the embryos of birds and mammals, which lack gills as adults. The first pharyngeal pouch is modified to form the tympanic cavity and the auditory (Eustachian) tube. The first pharyngeal groove forms the external auditory meatus. The membrane between the mandibular (first) and hyoid (second) arches is invaded by mesoderm and forms the tympanic membrane. The other pouches persist in the form of certain highly modified structures, retaining little or none of their original character. If a pharyngeal pouch fails to degenerate in the normal way, it may lead to the formation in the adult stage of a branchial fistula, in which there is an opening in the neck region that communicates with the pharynx.[101]

The presence of pharyngeal arches that do not develop into gills in birds and mammals can be best explained as due to evolution from gill-breathing, aquatic ancestors.[102] Darwin realized this when he wrote, "The points of structure, in which the embryos of widely different animals within the same class resemble each other, often have no direct relation to their conditions of existence. We cannot, for instance, suppose that in the embryos of the vertebrata the peculiar loop-like courses of the arteries near the branchial slits are related to similar conditions,—in the young mammal which is nourished in the womb of its mother, in the egg of the bird which is hatched in a nest, and in the spawn of a frog under water…[I]t is highly probable that with many animals the embryonic or larval stages show us, more or less completely, the condition of the progenitor of the whole group in its adult state...So again it is probable, from what we know of the embryos of mammals, birds, fishes, and reptiles, that these animals are the modified descendants of some ancient progenitor, which was furnished in its adult state with branchiæ, a swim-bladder, four fin-like limbs, and a long tail, all fitted for an aquatic life."[103]

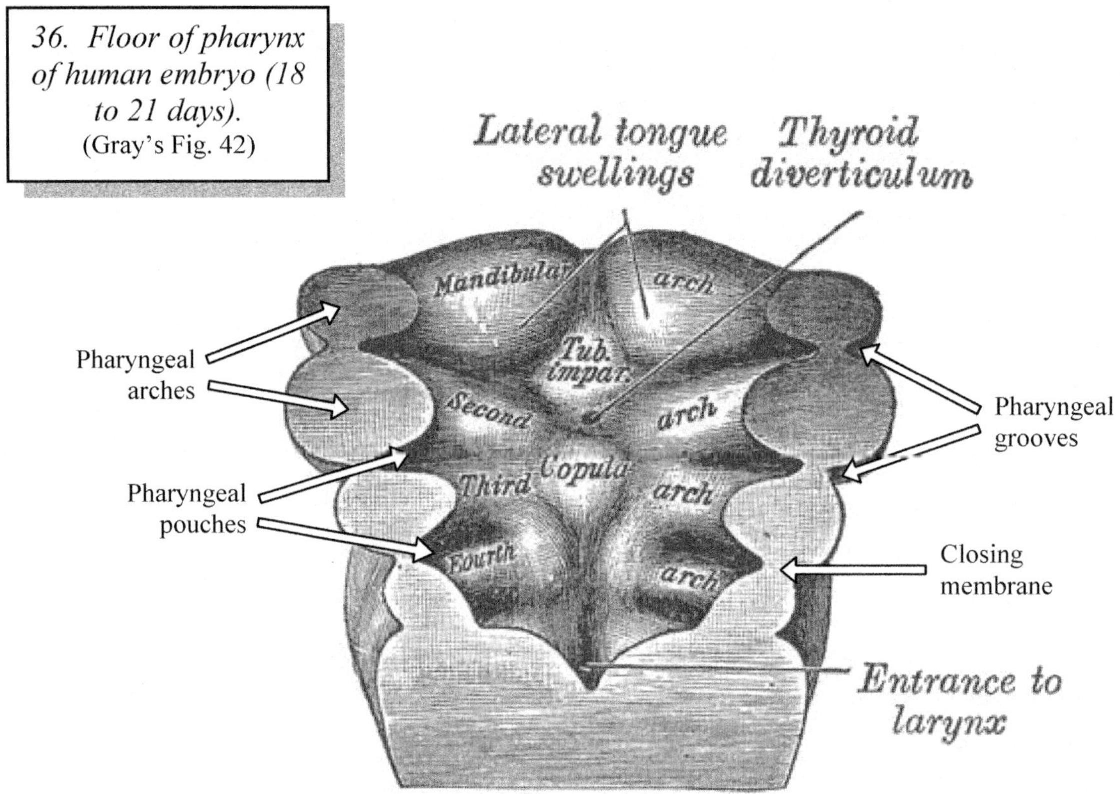

36. Floor of pharynx of human embryo (18 to 21 days). (Gray's Fig. 42)

Within the chordates, the presence of extraembryonic membranes in modern reptiles, birds, and mammals, including humans, can be best explained as due to descent of these groups from a common vertebrate ancestor. All reptiles have eggs (usually shelled) that contain a food source (the yolk) and a set of four membranes: yolk sac, amnion, allantois, and chorion. The yolk sac, which contains blood vessels that connect to the gut of the embryo, surrounds the yolk. It digests the yolk and transports nutrients to the embryo. The amnion, after which the egg is named (amniotic egg), encloses the embryo within a fluid-filled amniotic cavity. It is a sort of "portable pond" that enabled reptiles to become fully terrestrial. The allantois encloses a cavity that sequesters nitrogenous wastes excreted by the embryo. Finally, the chorion is the outermost membrane, enclosing all the other membranes and the embryo. It is permeable to gases but prevents the loss of water from the egg.

Mammals, including humans, also have these same four membranes during embryonic development, but usually with modified functions. The yolk sac does not surround any yolk, except in the mammals known as monotremes and marsupials, which lack a placenta. In most placental mammals, the yolk sac contributes to the formation of the digestive tract and produces the first blood cells and germ-line cells. The amnion still encloses the embryo in a cavity filled with amniotic fluid for protection and support. The blood vessels in the allantois become the umbilical blood vessels. The umbilical cord carries blood between the fetus and placenta. Lastly, the chorion forms the fetal portion of the placenta, which is normally fully formed and functional by the end of embryonic growth. The exchange of nutrient and waste molecules between maternal blood and fetal blood occurs in the placenta.

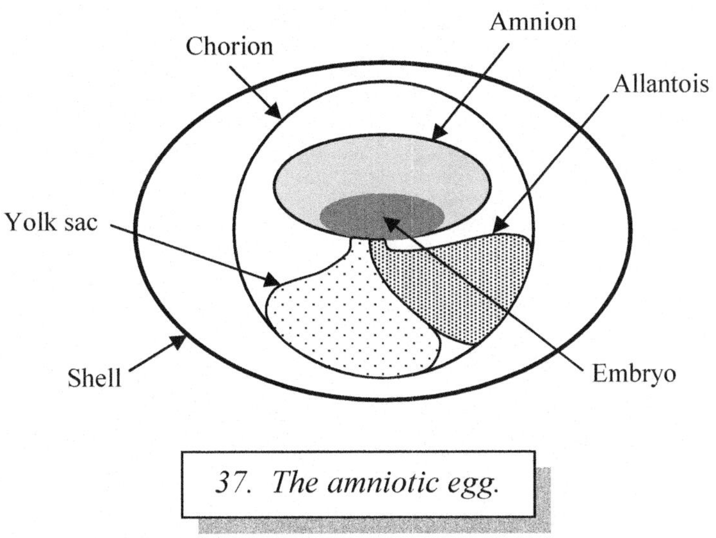

37. *The amniotic egg.*

Additional evidence for the common ancestry of reptiles, birds, and mammals includes the fact that monotremes, such as the platypus and the echidna, lay shelled eggs containing yolk, just like reptiles and birds do. Also, the embryos of marsupials obtain some nourishment from yolk and during the first two-thirds of prenatal development are encased in a thin shell. However, the shell disappears and the young are born live. Interestingly, although they do not need to hack through a hard egg-shell, several marsupial neonates (such as baby Brushtail possums, koalas, and bandicoots) even have a vestigial caruncle![xiv, 104, 105]

TOPIC 23. THE EVOLUTION OF HUMAN BIRTH

The much larger brain of humans (compared to other primates) is an adaptation that permits speech and culture, but requires a dramatically larger skull. However, bipedal locomotion becomes more inefficient as the pelvis becomes wider to accommodate the

---

[xiv]   The caruncle is an extension of the upper jaw that falls off soon after hatching, used to break open the shell; also called an egg-tooth.

head of the fetus. This can be easily observed by comparing how differently men and women run! Thus, birth in humans has evolved as an adaptive compromise between the need for a narrow pelvis and the need for a large brain.

The pelvic inlet of non-human primates, which usually walk quadrupedally (on four limbs), is usually large. Especially in brachiators (primates that swing through trees), the head (the largest part) of the fetus usually has plenty of room to pass through the birth canal. Labor is relatively easy and short, and many primates give birth in isolation and do not require assistance from others. The infant's head usually emerges in the **posterior position** (the back of the skull against the mother's sacrum and the face upward). The mother can pull the infant out without help, or the infant may pull itself out.

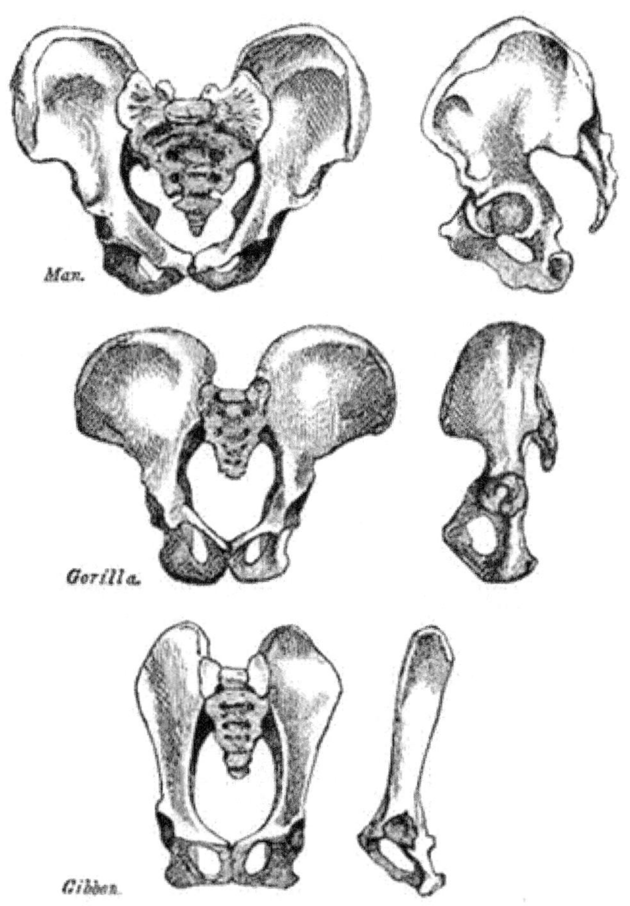

38. *Front and side views of the bony pelvis of Man, the Gorilla and Gibbon: reduced from drawings and from nature, of the same absolute length, by Mr. Waterhouse Hawkins.*
(Huxley, *The Relations of Man*, Fig. 16)

The narrowed pelvic inlet of humans is an adaptation that permits bipedal locomotion, but it is barely large enough for the infant's head. Since the pelvic inlet is broadest transversely but the outlet is widest sagittally, the large head of the relatively large-bodied human neonate has to rotate as it moves through the cervix and the vagina.[106] Even after such maneuvering, the skulls of infants often appear squashed for a few days following birth. Non-rotation is a common reason for failure to progress in labor and the subsequent emergency cesarean section. Labor is longer and more painful than in other primates, and human birth is difficult and risky for the vast majority of mothers and infants. The infant's head usually emerges in the **anterior position** (the face against the mother's sacrum and the skull against the pubis). The mother cannot pull the infant out without the risk of severely injuring its spinal cord and neck, and she therefore usually needs assistance to extract the infant. Human childbirth would be far better if the vagina opened outside the ring of pelvic bone, perhaps above the pubis on the lower abdomen.[107]

Since the female pelvis is probably as large as is feasible, a human infant is born with a brain that is much less developed than that of other primate infants. In most other mammals, the brain is nearly complete at birth. Other primates extend brain development into early postnatal growth. At birth, the brain of a rhesus (macaque) monkey (*Macaca mulatta*) is 65 percent of its adult size, a chimpanzee's is 40.5 percent, but the human's is only 23 percent. Chimps and gorillas attain 70 percent of adult brain size early in their first year, but humans do not do so until early in their third year. Humans are unique among mammals in the extent to which the brain keeps growing well after birth.
In addition, a human infant is born with a skeleton that is much less calcified than that of other primate infants. Though humans grow *in utero* to larger sizes than any other primate, human skeletal development is retarded compared to that of any monkey or ape for which information is available.[108] In most other mammals, the skull is fully ossified at birth. In humans, the sutures between the skull bones do not fully close until well after adulthood. This allows the brain to continue its dramatic postnatal expansion. Only in humans are the ends of long bones and digits still entirely cartilaginous at birth. In fact, ossification centers are usually entirely absent in the phalanges of human neonates. Macaque monkeys, including the rhesus, attain this level of ossification at the eighteenth fetal week. When macaques are born at 24 weeks, their limb bones have attained a level of ossification not reached by humans until years after birth.

If humans were born at the same stage of development as that of other primates, gestation in humans would last an estimated 16 or 17 to 21 months, instead of the normal nine. As Stephen Jay Gould has observed, "Human babies are born as embryos, and embryos they remain for about the first nine months of life"[109] They can neither stand up nor in any way fend for themselves for a long time. Thus, human infants require extensive care after birth for a much longer period than other mammals, including apes. For example, human infants continue to have all food brought to them by their parents even after weaning, whereas weaned apes gather their own food.[110]

TOPIC 24. AN OVERVIEW OF THE NERVOUS AND ENDOCRINE SYSTEMS

To maintain homeostasis, the nervous and endocrine systems together coordinate the functions of all body systems. They are both means of internal communication. However, they are complementary, not redundant – that is, they do not duplicate each other's function.

The nervous system controls homeostasis via nerve impulses. Nerve impulses trigger the release of neurotransmitter molecules. A neurotransmitter is a chemical messenger that diffuses over short distances to excite or inhibit only nearby neurons, muscle cells, or gland cells. Nerve impulses typically produce their effects within several milliseconds and the effects are relatively brief.

The endocrine system controls homeostasis by secreting hormones into the interstitial fluid, which then diffuse into the nearby bloodstream. The bloodstream delivers the hormones to virtually all body cells, although a hormone produces a response only in certain target cells that possess receptors for it. Thus, a hormone is a chemical messenger that is secreted into the bloodstream and affects the metabolism of target cells usually located in another organ. Hormones may take seconds to hours to produce their effects and these effects are generally longer in duration than those of neurotransmitters.

The nervous and endocrine systems are so intimately related that they are sometimes difficult to distinguish. Neurons often stimulate or inhibit the release of hormones, and hormones often promote or inhibit the initiation of nerve impulses. Some neurons actually release hormones, such as oxytocin, and some endocrine cells are modified neurons. Some neurotransmitters and hormones produce identical effects on the same target cells, such as norepinephrine and glucagon that each causes the breakdown of glycogen by the liver. In some cases, a single chemical, such as norepinephrine or dopamine, may serve as both a neurotransmitter and hormone. Hence, the two systems together are sometimes regarded as a coordinated supersystem called the neuroendocrine system.

TOPIC 24.1. THE ENDOCRINE SYSTEM

The endocrine system consists of the endocrine glands and tissues of the body, including the hypothalamus (a part of the brain), pituitary gland, thyroid gland, parathyroid glands, adrenal glands, pancreas, gonads, pineal gland, and thymus gland. However, some glands, such as the pancreas, have both endocrine and exocrine functions. Also, hormones are produced by endocrine cells in many organs that are not usually considered to be glands – for example, the brain, heart, stomach, small intestine, and placenta.

Target cells for a particular hormone are those cells that have the appropriate receptor molecule that can bind to the hormone. The number of receptor molecules in a target cell may increase or decrease in order to increase or decrease, respectively, the sensitivity of that target cell to a particular hormone. The amount of hormone secreted by endocrine cells is regulated according to the body's need for the hormone at any given time.

There is no "master control center" that regulates the entire endocrine system, but the hypothalamus is the major integrating link between the nervous and endocrine systems and receives input from numerous sources. It forms the floor and walls of the third ventricle of the brain. In addition to many other functions, it exerts control over the pituitary gland and is itself a crucial endocrine gland. The hypothalamus and pituitary gland together play important roles in the regulation of virtually all aspects of growth, development, metabolism, and homeostasis.

The pituitary gland (or hypophysis) is about 1.3 cm in diameter and is located in the hypophyseal fossa of the sphenoid bone. The infundibulum of the pituitary gland attaches it to the hypothalamus. In humans, the pituitary gland consists of two anatomically and functionally separate parts: the adenohypophysis and the neurohypophysis.

The adenohypophysis forms 75 percent of the gland by weight and contains many glandular epithelial cells. It consists of two parts: a large anterior lobe (pars distalis) and the pars tuberalis, a small mass of cells adhering to the anterior side of the infundibulum. The pituitary gland in some animals has a region between the anterior lobe and the neurohypophysis called the pars intermedia (intermediate lobe), but in humans its cells mingle with those of the anterior lobe during fetal development and no longer form a separate lobe in the adult. The secretion of anterior lobe hormones is regulated by releasing hormones and inhibiting hormones that are secreted by the hypothalamus and carried to the anterior lobe by a complex of blood vessels called the hypothalamo-hypophyseal portal system. Blood from the superior hypophyseal arteries enters a capillary bed, called the primary plexus of the portal system, in the hypothalamus. From here, blood drains into the hypophyseal portal veins that pass down the surface of the infundibulum to another capillary network, called the secondary plexus of the portal system, in the anterior lobe of the pituitary. From here, blood drains into the anterior hypophyseal veins and eventually returns to the heart. The anterior lobe secretes seven major hormones:

- Human growth hormone (hGH), or somatotropin, stimulates general body growth and regulates certain aspects of metabolism.
- Prolactin (PRL) stimulates milk production.
- Adrenocorticotropic hormone (ACTH), or corticotropin, stimulates the adrenal cortex to secrete hormones.
- Melanocyte-stimulating hormone (MSH) affects skin pigmentation.
- Thyroid-stimulating hormone (TSH), or thyrotropin, regulates the activities of the thyroid gland.
- Follicle-stimulating hormone (FSH) initiates development of oocytes and stimulates production of sperm.
- Luteinizing hormone (LH) stimulates ovulation and secretion of testosterone from the testes.

Notice that four of the anterior pituitary hormones are tropins or tropic hormones, that is, hormones that influence another endocrine gland or tissue. They include TSH, ACTH, FSH, and LH. Since FSH and LH regulate the functions of the gonads, they are called gonadotropins.

The neurohypophysis comprises the posterior one quarter of the pituitary. It consists of two parts: the infundibulum, also known as the pituitary stalk and the posterior lobe (pars nervosa), which makes up most of the neurohypophysis. It contains axons and axon terminals of neurons (neurosecretory cells) whose cell bodies lie in the hypothalamus. The posterior lobe hormones are produced in the cell bodies of the neurosecretory cells in the hypothalamus. The axons of these cells form the hypothalamo-hypophyseal tract in the infundibulum and carry these hormones to axon terminals in the posterior lobe. Nerve impulses that travel to the axon terminals trigger the release of the posterior lobe hormones into nearby capillaries, called the capillary plexus of the infundibular process. These capillaries receive blood from the inferior hypophyseal arteries and drain into the posterior hypophyseal veins. The posterior lobe releases two major hormones:

- Oxytocin is involved in labor and milk ejection.
- Antidiuretic hormone (ADH), or vasopressin, decreases urine volume and increases blood pressure.

The thyroid gland is located just below the larynx. It consists of a right lateral lobe and a left lateral lobe, which are located on either side of the trachea. The two lobes are connected by an isthmus that lies in front of the upper end of the trachea. Sometimes, the thyroid has a small pyramid-shaped lobe that extends upward from the isthmus. Three hormones are secreted by the thyroid. Follicular cells produce the two so-called thyroid hormones – thyroxine, also called tetraiodothyronine (T_4), and triiodothyronine (T_3) – that regulate the rate of metabolism, growth, and development. Parafollicular cells, or C cells, produce the hormone calcitonin, which influences calcium homeostasis by decreasing blood calcium concentration.

The tiny parathyroid glands are partially embedded in the posterior surface of the thyroid. There are usually four – a superior pair and an inferior pair. Their so-called chief cells secrete parathyroid hormone (PTH), or parathormone. PTH increases blood calcium and magnesium levels, decreases blood phosphate levels, and promotes the formation of calcitriol, which is the active form of vitamin D.

The adrenal (suprarenal) glands are located superior to the two kidneys in the retroperitoneal space. Each adrenal gland has two physiologically independent regions: the outer cortex and the inner medulla.

The adrenal cortex has three subdivisions: the outer zona glomerulosa, which secretes mineralocorticoids, mainly aldosterone, that affect levels of sodium and potassium in the blood; the middle zona fasciculata, which secretes glucocorticoids, mainly cortisol, that affect glucose metabolism; and the inner zona reticularis, which secretes minute amounts of weak androgens, mainly testosterone, that have masculinizing effects. The cortical hormones provide a long-term response to stress and are essential for life. Their release is controlled mainly by ACTH.

The adrenal medulla secretes two hormones: epinephrine (adrenaline) and norepinephrine (noradrenaline). Its hormones provide a short-term response to stress, such as an emergency situation. The medullary hormones are responsible for the fight-or-flight response and are not essential for life. Their release is controlled by autonomic nerve impulses from the hypothalamus.

The pancreas is a flattened organ located retroperitoneally, posterior and slightly inferior to the stomach. About 99 percent of it is an exocrine digestive gland, but scattered among the exocrine acini are one to two million tiny endocrine cell clusters called pancreatic islets, or islets of Langerhans. The islets secrete several pancreatic hormones, the most important of which are insulin and glucagon. Insulin is produced by beta (β) cells in the pancreatic islets and acts to decrease the level of glucose in the blood. Glucagon is produced by alpha (α) cells and acts to increase the level of glucose in the blood.

Like the pancreas, the gonads are both endocrine and exocrine glands. Their exocrine product is eggs and sperm, and their endocrine product is the gonadal hormones, most of which are steroids. The ovaries are paired oval structures located in the pelvic cavity. They secrete: the hormones estrogen and progesterone, which are responsible for the development and maintenance of female sexual characteristics; inhibin, which inhibits secretion of FSH from the anterior pituitary; and relaxin, which relaxes the pubic symphysis and helps dilate the cervix just prior to childbirth. The testes are paired oval structures located in the scrotum. The testicular interstitial (Leydig) cells of the testis secrete the hormone testosterone, which regulates production of sperm and stimulates growth and development of male sexual characteristics. The testicular sustentacular (Sertoli) cells of the testis secrete inhibin, which inhibits secretion of FSH from the anterior pituitary and consequently stabilizes the rate of sperm production.

Shaped like a pine cone, the pineal gland (epiphysis cerebri) is attached to the roof of the brain's third ventricle, beneath the posterior end of the corpus callosum. The philosopher René Descartes thought it was the seat of the human soul. If so, children must have more soul than adults – a child's pineal gland is about 8 mm long and 5 mm wide, but after age seven it regresses rapidly and is no more than a tiny shrunken mass of fibrous tissue in the adult. Although philosophers no longer look for the human soul in the pineal, its function is still unclear to scientists. It secretes melatonin, which plays a role in the setting of the body's biological clock and may affect human reproductive functions.

The thymus is located between the sternum and aortic arch in the mediastinum, superior to the heart. Like the pineal, it is larger in infants and children, but atrophies after puberty. In adults, the thymus is a shrunken mass of mostly fibrous and adipose tissue. Nevertheless, it plays a major role in immunity. If the thymus is removed from newborn mammals, they waste away and never develop immunity. The hormones it secretes include thymosin, thymic humoral factor (THF), thymic factor (TF), and thymopoietin. The thymic hormones promote the proliferation and maturation of T cells and may also retard the aging process.

In addition to the endocrine glands described above, there are other organs and tissues that secrete hormones. The enteroendocrine cells of the gastrointestinal tract secrete at least ten different enteric hormones, including gastrin, glucose-dependent insulinotropic peptide (GIP), secretin, and cholecystokinin. The placenta secretes human chorionic gonadotropin (hCG), estrogen, progesterone, and human chorionic somatomammotropin (hCS). The kidneys secrete erythropoietin (EPO), which stimulates the bone marrow to produce red blood cells, and calcitriol, which affects the handling of calcium by the kidneys, small intestines, and bones. The heart secretes atrial natriuretic peptide (ANP), which acts to lower the blood pressure, and adipose tissue secretes leptin, which suppresses appetite and may increase the activity of FSH and LH.

TOPIC 24.2. THE NERVOUS SYSTEM: AN INTRODUCTION

The nervous system has three major functions: receiving sensory input, performing integration (summing up the input), and stimulating motor output. Structurally, the nervous system consists of two major divisions: the central nervous system (CNS), which consists of the brain and spinal cord, and the peripheral nervous system (PNS), which is the part of the nervous system outside the brain and spinal cord. The PNS includes the cranial nerves, which emerge from the brain, and the spinal nerves, which emerge from the spinal cord.

The two types of cells in nervous tissue are neurons (nerve cells), which possess electrical excitability, and neuroglial cells (or glial cells), which support, protect, and nourish the neurons.

Each neuron has a cell body (or soma), which contains the nucleus and other organelles within the cytoplasm. Also, neurons have two types of processes (or neuron fibers) that contain cytoplasm and extend away from the cell body: dendrites and axons. A dendrite is a branched process that conducts nerve impulses toward the cell body, whereas an axon is a process that conducts nerve impulses away from the cell body. Most neurons have one or more dendrites, but each neuron has only one axon. An axon arises from a cone-shaped region of the cell body called the axon hillock and occasionally has side-branches called axon collaterals. The axon and its collaterals end at tiny structures called axon terminals, which contain synaptic vesicles that store molecules of neurotransmitter.

There are three functional types of neurons: sensory, motor, and interneurons. Sensory (or afferent) neurons carry nerve impulses to the CNS from sensory receptors. Motor (or efferent) neurons carry nerve impulses away from the CNS to muscles or glands, which are called effectors. Interneurons (or association neurons) are found only within the CNS and carry nerve impulses between various parts of the brain and spinal cord.

The structural classification of neurons is based on the number of processes that extend from the cell body. Multipolar neurons, which comprise most of the neurons in the CNS, have dendrites and one axon. Bipolar neurons, which are found in the retina, inner ear, and olfactory area of the brain, have one dendrite and one axon. Unipolar neurons, which are always sensory neurons, have just one process extending from the cell body. A short distance from the cell body, this process divides into two branches that function as one axon: the branch associated with dendrites is called the peripheral process and the other branch that enters the central nervous system is called the central process (see Figure 39).

The junction between one neuron and another neuron or an effector cell is called a synapse. The synapse between a motor neuron and a muscle cell is called a neuromuscular junction (see Topic 13). The synapse between a motor neuron and a glandular cell is called a neuroglandular junction.

The neuroglial cells of the PNS consist of Schwann cells and satellite cells. Schwann cells often form a white myelin sheath consisting of layers of plasma membrane wrapped around an axon. Each cell produces part of the myelin sheath around a single axon and participates in regeneration of axons. Gaps between Schwann cells form the neurofibral nodes (nodes of Ranvier) that accelerate impulse conduction. Satellite cells surround the cell bodies of neurons, which they thereby support.

The neuroglial cells of the CNS consist of astrocytes, oligodendrocytes, microglial cells, and ependymal cells. Astrocytes have cytoplasmic processes that form bridges between neurons and capillaries, providing structural support. Oligodendrocytes produce the myelin sheath that covers axons, each cell myelinating portions of several adjacent axons. Microglial cells are small phagocytes that migrate to the site of an injury, where they engulf microbes and remove debris. Ependymal cells line cavities and secrete cerebrospinal fluid (CSF) that fills the cavities.

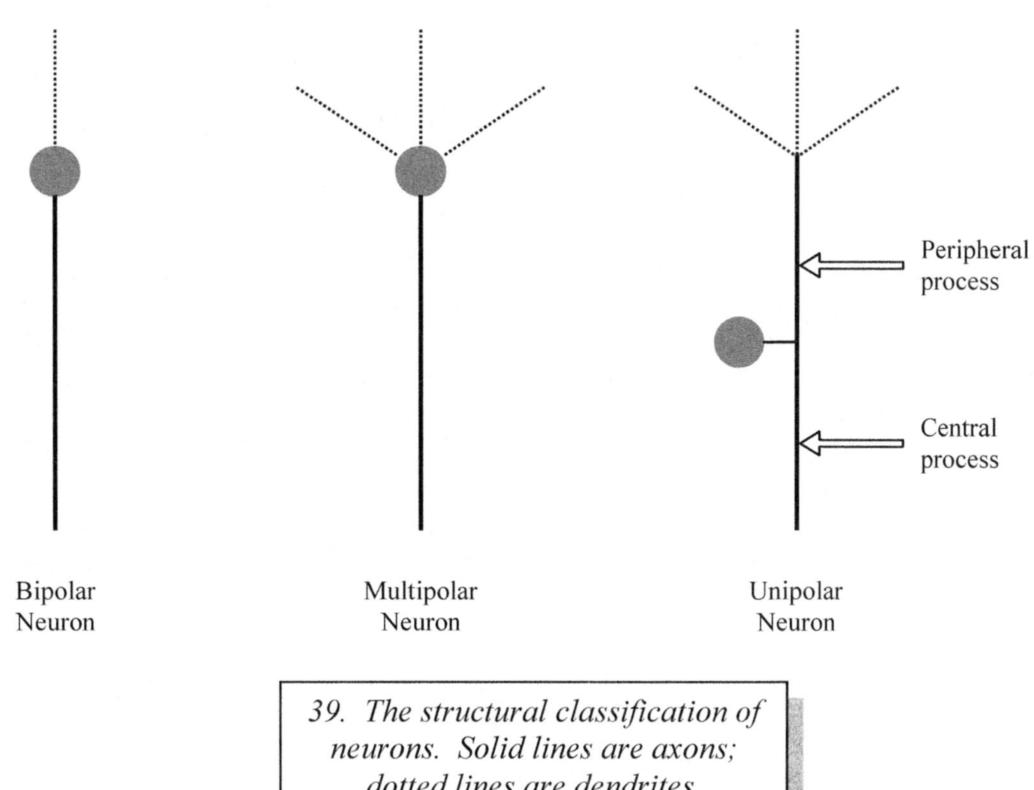

39. The structural classification of neurons. Solid lines are axons; dotted lines are dendrites.

The speed of a nerve impulse along an axon is determined by the diameter of the axon, the presence or absence of myelin, and temperature. Large fibers conduct impulses more rapidly than small ones because conduction occurs only along the plasma membrane of a fiber, not deep within the cytoplasm, and larger fibers have more surface area. Myelinated fibers conduct impulses faster than nonmyelinated ones.

So why aren't all nerve fibers large and myelinated? Apparently evolution did not result in this. Questions like this can be difficult to answer, because the evolutionary details are not available and the process is not open to experiment, but cost-benefit considerations can sometimes help.

The above question assumes that speed is always beneficial. However, this may not be the case. For example, high speed attack and escape behavior seems clearly to be an evolutionary advantage, yet there is also a need for slow, precise life-sustaining movements like feeding, grooming, and even stealthy attacks and escapes. Thus, both slow and fast muscles, muscle fibers, and muscle proteins have evolved. There would be little point in maximizing the speed of an axon that innervates a slow muscle fiber – unless that maximization could be without cost.

So what is the cost of speed? Myelination itself is one – numerous Schwann cells and oligodendrocytes are required to support the action of a single axon. Size is another. Myelinated axons are all large, and the myelination makes them larger. The space available to contain axons is not unlimited – nerves have to be able to fit inside the body parts they supply. So the cost of making all axons large and myelinated to achieve uniform maximal speed is a smaller number of axons. Fewer axons means poorer control of effectors and poorer sensation. Poor sensation can be just as big an evolutionary threat as low ground speed.

So while there **is** sometimes an evolutionary benefit to fast axons, there is also sometimes an evolutionary benefit to slow and space efficient axons – they allow for more parallel paths and richer information capabilities at an optimal speed. In addition, at least one report suggests that below a certain threshold in diameter, a nonmyelinated axon may be **faster** than its myelinated counterpart. The conduction velocity vs. axon diameter graph in Figure 40 is the result of a theoretical analysis by William A.H. Rushton.[111] The diameter of the myelinated fiber includes the myelin. Notice that myelination gives faster speeds only above about 1μm diameter. Below that diameter nonmyelinated axons are actually slightly faster. (This is partly because the proportion of space occupied by myelin starts to be very great in small axons.) Human anatomy reflects this constraint – myelinated axons less than 1μm are rare, and nonmyelinated axons larger than 1μm are similarly rare.

The PNS contains structures called ganglia (singular: ganglion) and nerves.

- A ganglion is a cluster of neuron cell bodies.
- A nerve is a cable-like organ containing axons arranged in parallel bundles and enclosed by connective tissue. Sensory nerves contain the axons of sensory

neurons, whereas motor nerves contain the axons of motor neurons. Mixed nerves, which are the most common, contain the axons of both sensory and motor neurons, for example, the spinal nerves. Nearly all nerves contain both myelinated and nonmyelinated axons.

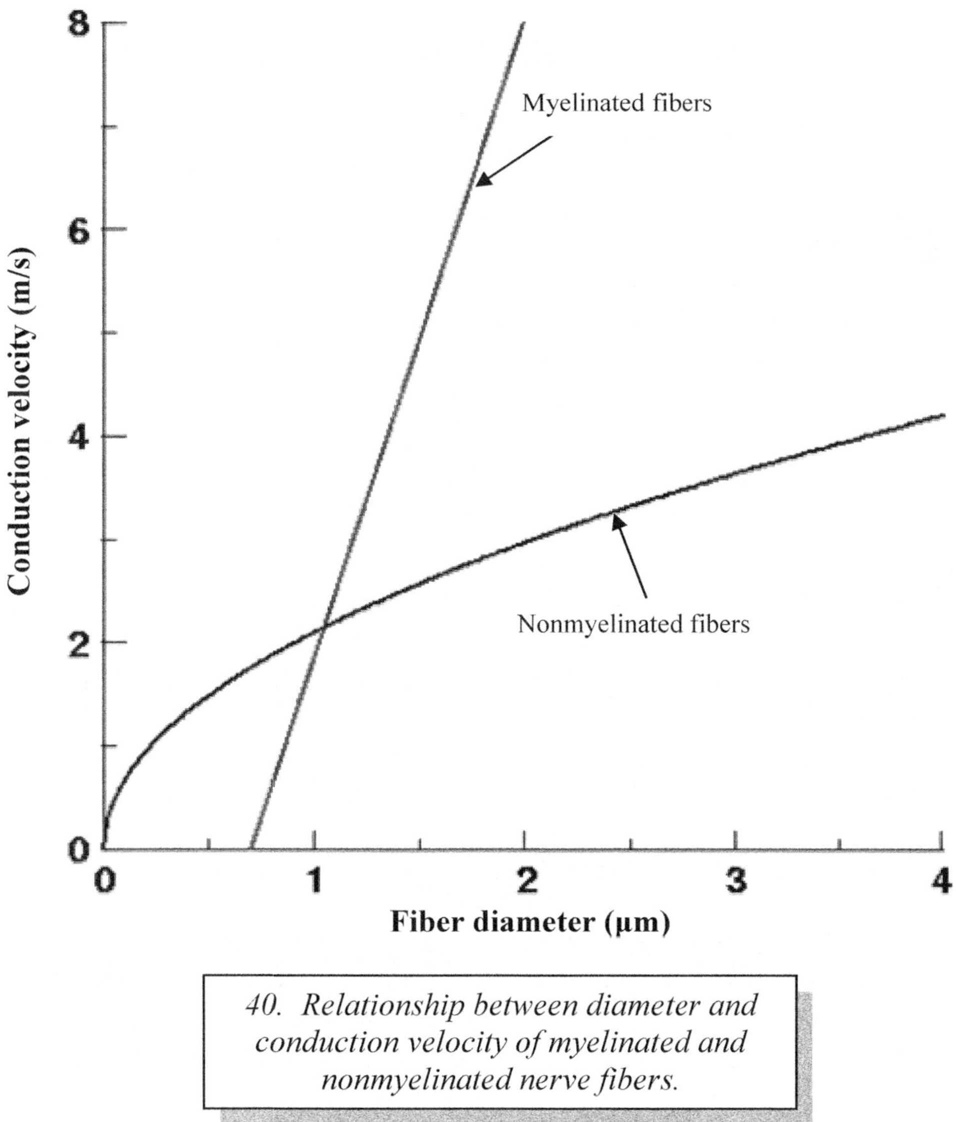

40. Relationship between diameter and conduction velocity of myelinated and nonmyelinated nerve fibers.

The CNS contains structures called nuclei and tracts.

- A nucleus is a cluster of neuron cell bodies.
- A tract is a bundle of axons having the same origin, destination, and function. Ascending tracts contain the axons of sensory neurons and descending tracts contain the axons of motor neurons. A commissure is a tract that crosses from one side of the CNS to the other.

The CNS has some regions that appear white and others that appear gray. White matter contains millions of axons, most of which are myelinated. The myelin sheaths give white matter its color. Gray matter contains neuron cell bodies, dendrites, axon terminals, nonmyelinated axons, neuroglial cells, and thus no myelin. The white matter of the spinal cord surrounds a core of gray matter. The white matter of the brain also surrounds centrally located gray matter but in two regions (the cerebrum and cerebellum) there is an additional layer of gray matter located superficially, called the cortex.

TOPIC 24.3. THE SPINAL CORD AND SPINAL NERVES

Both the brain and the spinal cord are surrounded and protected by bone, the meninges, and cerebrospinal fluid. Cranial bones form the cranial cavity that contains the brain, and vertebrae form the vertebral canal that contains the spinal cord. The cranial cavity and vertebral canal together form the dorsal body cavity. The meninges (singular: meninx) consist of three fibrous membranes that line the dorsal body cavity:

- The outer dura mater is dense irregular connective tissue that lies next to the skull and vertebrae. In the cranial cavity, it consists of two layers: an outer periosteal layer, equivalent to the periosteum of the cranial bone, and an inner meningeal layer. In some places, the two layers are separated by dural venous sinuses, which collect blood that has circulated through the brain and drain into the internal jugular veins of the neck. Also, in some places the meningeal layer of the dura folds inward to separate major parts of the brain. In the vertebral canal, the periosteal layer of the dura mater is absent and the meningeal layer forms a dural sheath that encloses the spinal cord. The space between the sheath and vertebrae is called the epidural space and is occupied by adipose and loose fibrous connective tissue for protection.
- The middle arachnoid membrane is a connective tissue covering with thin fibrous strands like a spider-web (hence the name) that attach to the pia mater. The space between the arachnoid and dura is called the subdural space.
- The inner pia mater is a layer of very thin loose connective tissue and closely follows the contours of the brain and spinal cord. It is vascularized to provide nutrients and oxygen to the spinal cord and brain. The space between the pia and the arachnoid is called the subarachnoid space. Denticulate ligaments are thickenings of the pia mater that help anchor the spinal cord in place.

Cerebrospinal fluid is a clear, colorless liquid that fills the subarachnoid space of the meninges, the ventricles (cavities) within the brain, and the central canal of the spinal cord, all of which are interconnected. Ependymal cells in the brain produce CSF from arterial blood plasma by filtration and secretion. They cover a network of blood capillaries, forming a structure called a choroid plexus[xv] on the wall of each ventricle. After circulating throughout the CNS, CSF is absorbed into venous blood of the dural sinuses. CSF supplies the CNS with nutrients and collects wastes. It also protects the brain from striking the cranium when the head is jolted, and maintains proper electrolyte

---

[xv]   The choroid plexus was so-named because it histologically resembles the chorion, one of the extraembryonic membranes that form during prenatal development.

balance for action potentials of neurons in the CNS. Tight junctions between ependymal cells form the blood-CSF barrier that prevents potentially harmful substances from leaking into the CSF from the blood.

The spinal cord extends from the foramen magnum through the vertebral canal of the dorsal body cavity to the level of the second lumbar vertebra (L2) in an adult.   It averages about 1.8 cm in diameter and ranges from 42 to 45 cm in length. Thus, the spinal cord occupies only the upper two-thirds of the vertebral canal.

A spinal segment is a section of the spinal cord from which a pair of spinal nerves emerge, one on each side. There are a total of 31 pairs of spinal nerves that are named and numbered according to the region and level of the vertebral canal from which they emerge:

- Eight pairs of cervical spinal nerves (the first pair emerges between the atlas and the occipital bone).
- Twelve pairs of thoracic spinal nerves.
- Five pairs of lumbar spinal nerves.
- Five pairs of sacral spinal nerves.
- One pair of coccygeal spinal nerves.

The spinal cord has two enlargements where nerves supplying the limbs emerge: the cervical enlargement that supplies the upper extremities and the lumbar enlargement that supplies the lower extremities. Inferior to the lumbar enlargement, the spinal cord tapers to a point called the conus medullaris at the superior border of vertebra L2. The canal of vertebrae L2 to S5 contains a thick bundle, called the cauda equina, consisting of spinal nerve roots that arise from the lumbar enlargement and conus medullaris. Thus, these spinal nerves are unusual in that they do not emerge from the vertebral canal at the same level from which they leave the spinal cord.

A transverse section of the spinal cord reveals longitudinal grooves on its ventral and dorsal sides: the anterior median fissure and the posterior median sulcus, respectively. In this view, the gray matter of the spinal cord appears in a central **H**-shaped region that is surrounded by bundles of white matter. Each side of the letter "H" consists of a dorsal (posterior) horn extending toward the dorsolateral surface of the cord and a ventral (anterior) horn extending toward the ventrolateral surface. In the thoracic, upper lumbar, and sacral segments of the spinal cord, an additional lateral gray horn is adjacent to each junction of a dorsal horn and ventral horn. The right and left regions of gray matter are connected by the gray commissure, which crosses the middle of the cord. The central canal is at the center of the gray commissure and extends throughout the entire length of the spinal cord.

The white matter of the spinal cord is divided by the ventral and dorsal gray horns into regions called funiculi (singular: funiculus) or columns: an anterior (ventral) column, lateral column, and posterior (dorsal) column on each side. Each column is subdivided into tracts or fasciculi (singular: fasciculus). Spinal cord tracts are continuous with tracts in the brain. Ascending (sensory) tracts transmit nerve impulses upward to the brain. The names of most ascending tracts consist of the prefix "spino-" followed by a word root denoting the destination of the tract. The main sensory tracts on each side of the cord are the posterior columns, which consist of the fasciculus gracilis and the fasciculus cuneatus, and the spinothalamic tract, which is found in the anterior and lateral columns. Descending (motor) tracts transmit nerve impulses downward from the brain. The names of most descending tracts consist of a word root denoting the origin of the tract in the brain followed by the suffix "-spinal". The main motor tracts on each side of the cord are the corticospinal tracts, reticulospinal tracts, tectospinal tract, and the vestibulospinal tract. The white matter of the right and left sides of the spinal cord are connected by the anterior white commissure, which is anterior to the gray commissure.

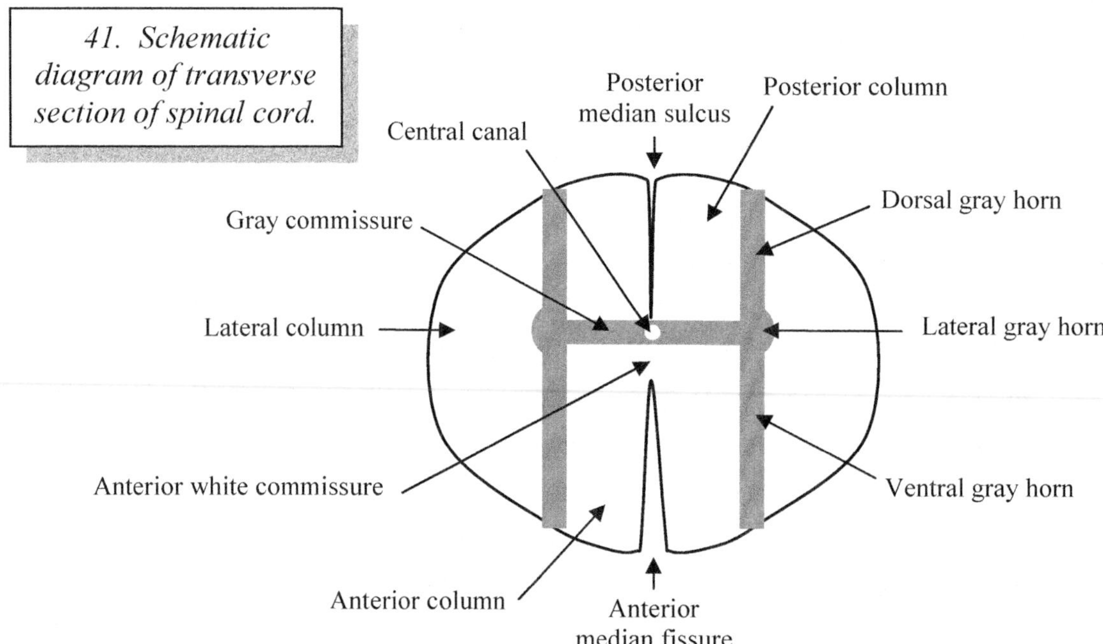

41. Schematic diagram of transverse section of spinal cord.

The different spinal cord segments vary in size, shape, relative amounts of gray and white matter, and precise distribution and shape of gray matter. Each spinal nerve emerges from a spinal segment by two short branches or roots, which lie within the vertebral column. The dorsal root carries the axons of sensory neurons whose cell bodies are located in the dorsal root ganglia. The ventral root carries the axons of motor neurons whose cell bodies are located in the horns of the spinal cord. Cell bodies of motor neurons supplying skeletal muscles are located in the ventral gray horns, while cell bodies of motor neurons supplying smooth muscle, cardiac muscle, or glands are located in the lateral gray horns. The dorsal and ventral roots converge to form a spinal nerve at the intervertebral foramen, so that every spinal nerve is a mixed nerve. Shortly after passing through its intervertebral foramen, a spinal nerve divides into three or four branches:

- The dorsal ramus (plural: rami) innervates the muscles and joints in that region of the spine and the skin of the back.

- The ventral ramus innervates the ventral and lateral skin and muscles of the trunk and gives rise to nerves of the extremities.
- The meningeal branch reenters the vertebral canal through the intervertebral foramen and innervates the meninges, vertebrae, and spinal ligaments.
- The spinal nerves in the thoracic and lumbar regions have a fourth branch known as the visceral branch, which is part of the autonomic nervous system.

The spinal cord has three main functions. First, it is a conduit for nerve impulses to and from the brain. Second, the spinal cord serves as an integrating center for spinal reflexes. Reflexes are fast, predictable, involuntary responses by peripheral effectors to stimulation. Examples include the swallowing reflex, the blinking reflex, and the patellar (knee-jerk) reflex. The pathway into, through, and out of the CNS followed by a nerve impulse that produces a reflex response is called a reflex arc. It consists of five functional parts:

- A sensory receptor (distal end of a sensory neuron or an associated sensory structure).
- A sensory neuron (transmits nerve impulses to the integration site in the spinal cord or brainstem).
- An integrating center (relays nerve impulses from a sensory to a motor neuron).
- A motor neuron (transmits nerve impulses to the effector).
- An effector (muscle or gland that responds to motor nerve impulses).

A spinal reflex occurs through an integrating center in the spinal cord and involves spinal nerves, while a cranial reflex occurs through an integrating center in the brainstem and involves cranial nerves. A reflex arc is monosynaptic if the sensory neuron and motor neuron synapse directly at the integrating center and it is polysynaptic if there are one or more interneurons at the integrating center. A reflex is a somatic reflex if the effector is a skeletal muscle, while it is an autonomic (visceral) reflex if the effector is smooth muscle, cardiac muscle, or a gland.

A third major function of the spinal cord is mediated by reverberating circuits called central pattern generators (CPGs) found in its gray matter. They produce the necessary sequence of motor outputs to the extensor and flexor muscles of the legs to enable walking. In the body, CPGs are turned on and off by voluntary motor control in the brain.

TOPIC 24.4. THE BRAIN AND CRANIAL NERVES

The average adult brain weighs about 1,600 g (3.5 lb) in men and 1,450 g in women – its size is proportional to body size, not intelligence. Neanderthals on average had larger brains than modern humans because they had larger bodies. For comparison, the average brain of an adult sperm whale weighs about 7,800 g.

The brain consists of four major regions: the brainstem, the diencephalon, the cerebellum, and the cerebrum. The cerebellum and cerebrum each consist of two hemispheres. An important feature of the brain consists of four CSF-filled cavities called ventricles. The two lateral ventricles are found in the left and right cerebral hemispheres. The so-called third ventricle is located in the diencephalon, connected to the lateral ventricles by the interventricular foramina. The so-called fourth ventricle is located between the brainstem and the cerebellum; it is connected to the third ventricle by the cerebral aqueduct and to the subarachnoid space of the meninges.

The brain receives most of its blood from the blood vessels of the cerebral arterial circle. Tight junctions between endothelial cells of brain capillaries help form the blood-brain barrier that prevents substances from leaking into brain tissue from the blood. A few small regions of the brain, such as the pineal gland and part of the hypothalamus, lack the blood-brain barrier and can monitor chemical changes in the blood.

The brainstem consists of three parts: the medulla oblongata, the pons, and the midbrain. The medulla oblongata is the lowest portion of the brainstem and is anterior to the cerebellum. Its gray matter contains discrete nuclei that are vital centers for regulating heartbeat, breathing, and blood pressure. The medulla also contains all ascending and descending tracts that connect the spinal cord to the brain. The medullary pyramids are bulges on the anterior surface of the medulla containing descending tracts called the corticospinal tracts. Most tracts decussate (cross over) as they pass through the medulla and as a result, the right side of the brain controls the left side of the body, and vice versa. The pons is superior to the medulla oblongata, inferior to the midbrain, and anterior to the cerebellum. Its gray matter contains nuclei concerned with sleep, posture, respiration, swallowing, and bladder control. The pons also contains tracts that connect the cerebrum with the cerebellum, or are part of the ascending and descending tracts that bridge the spinal cord to the brain. The midbrain extends upward from the pons to the lower portion of the diencephalon. It encloses the cerebral aqueduct and contains tracts passing between the cerebrum and the spinal cord or cerebellum. Major functions of the midbrain include: receiving and integrating auditory information, coordinating visual reflexes, and sending sensory information to higher brain centers.

The diencephalon is the portion of the brain in the region of the third ventricle, superior to the midbrain, and consists of three major parts: the thalamus, the hypothalamus, and the epithalamus. The thalamus lies in the lateral walls of the third ventricle and makes up about 80 percent of the diencephalon. The thalamus may be regarded as a "gateway to the cerebral cortex." It receives sensory (except olfactory) impulses from all parts of the body and relays them to the cerebrum. Also, the gray matter of the thalamus contains nuclei concerned with movement planning and control. The hypothalamus forms the floor of the third ventricle. Composed of about a dozen nuclei, it helps control the functioning of most internal organs and controls the secretions of the pituitary. The epithalamus lies in the roof of the third ventricle and consists of the pineal gland and the habenular nuclei, the latter of which is involved in olfaction.

The cerebellum, the second-largest part of the brain, is inferior to the posterior portion of the cerebrum, posterior to the brainstem, and separated from the latter by the fourth ventricle. The right and left hemispheres of the cerebellum are joined by a constricted median portion called the vermis. The cerebellum has a cortex of gray matter, beneath which is white matter arranged in tracts called arbor vitae, named after their resemblance to the branches of a tree. Even deeper, within the white matter, are the cerebellar nuclei. The cerebellum coordinates contractions of skeletal muscles and maintains normal muscle tone, posture, and balance.

The cerebrum is the largest and most superior part of the brain, and is alone responsible for consciousness. It consists of three major parts: an outer cortex of gray matter, underlying white matter, and discrete masses of gray matter at the base of the cerebrum (basal nuclei[xvi]). The cerebral cortex, which is about 2 to 4 mm deep, contains most of the gray matter. The white matter constitutes most of the cerebral volume, but is not a decision-making (integrative) center. The cerebral white matter consists of three types of tracts:

- Projection tracts extend vertically to lower brain or spinal cord destinations and conduct impulses between the cerebrum and the peripheral nervous system.
- Commissural tracts cross horizontally from one cerebral hemisphere to the other and conduct impulses between them. Most of these tracts pass through the large **C**-shaped corpus callosum, enabling the two hemispheres to "communicate."
- Association tracts horizontally connect different regions within the same hemisphere – they do **not** extend from one side of the brain to the other.

The cerebral cortex is rolled upon itself, forming folds separated by grooves. The folds of cortex are called gyri (singular: gyrus) or convolutions. The deepest grooves are called fissures and the shallower grooves are called sulci (singular: sulcus). As a result of its gyri, the cerebral cortex makes up 40 percent of the mass of the entire brain and has a surface area of about 2,500 cm^2 – its area would be only one-third of this if it had a smooth surface. The longitudinal fissure separates the cerebrum into right and left hemispheres, which remain connected internally by a mass of white matter called the corpus callosum.

The cortex of each cerebral hemisphere is divided into lobes distinctly bordered by certain fissures and unusually deep sulci. These lobes are visible superficially and are named after the bones of the skull which each lies beneath: the frontal, parietal, temporal, and occipital. The central sulcus separates the frontal lobe from the parietal lobe. Located just anterior to the central sulcus in the precentral gyrus is the primary motor area of the cerebral cortex, which is the major control region for initiation of voluntary movements. Located just posterior to the central sulcus in the postcentral gyrus is the primary somatosensory area of the cerebral cortex, which receives general sensory information pertaining to touch, temperature, pain, pressure, and proprioception. About 90 percent of the cortex is a six-layered tissue called the neocortex; the remainder of the cortex is not stratified. The limbic system[112] (or so-called limbic lobe) actually consists of parts of other (frontal, temporal, and parietal) lobes. This system is involved in

---

[xvi]   They are often called basal ganglia, but the term "ganglion' is best reserved for clusters of neuron cell bodies in the PNS.

olfaction, emotions, learning, and memory.

The basal nuclei located in each cerebral hemisphere help control large, automatic movements of skeletal muscles and help regulate muscle tone. There are five basal nuclei in each cerebral hemisphere, lateral to the thalamus: the caudate nucleus, putamen, globus pallidus, amygdala, and claustrum. The putamen and globus pallidus together are called the lentiform nucleus. The basal nuclei are part of the extrapyramidal system to be discussed later.

Just like the anatomy of any other organ or organism, the anatomy of the brain becomes much clearer and more meaningful when you examine it in light of the evolutionary processes that created it.[113] The various species of vertebrates are very similar in the way that their brains are organized. For example, all vertebrates have a forebrain (cerebrum and diencephalon), midbrain, and hindbrain (cerebellum, medulla oblongata, and pons), within which are found all the major neural systems that have evolved to perform functions common to all species. However, the various species also have areas of the brain that have specialized in distinctive ways in response to the specific constraints of their environments. The model of the triune brain first proposed by Paul MacLean (1913-2007) in 1970 is a useful piece of shorthand for the complex evolutionary history of the human brain.[114] According to this idea, the human brain evolved in three stages, resulting in three divisions that co-inhabit the human skull:

- The **reptilian brain**, or R-complex, first appeared in fish, nearly 500 million years ago. It continued to develop in amphibians and reached its most advanced stage in reptiles, roughly 250 million years ago. Our reptilian brain includes the main structures found in a reptile's brain: the brainstem and the cerebellum. It also includes the oldest basal nuclei: the globus pallidus and the olfactory bulbs. The reptilian brain has the same type of archaic behavioral programs as snakes and lizards. It is rigid, obsessive, compulsive, ritualistic, paranoid, and is said to be "filled with ancestral memories". It keeps repeating the same behaviors over and over again, never learning from past mistakes. This part of our brain controls muscles, balance, and autonomic functions, such as breathing and heartbeat. Thus, the reptilian brain is active even in deep sleep.
- The **limbic system**, or paleomammalian brain, became prominent in small mammals, about 150 million years ago. In 1952, MacLean first coined the name "limbic system" for this part of the brain. It corresponds to the brain of most mammals, especially those that evolved earlier. The main structures of the limbic system are the hippocampus, the amygdala, and the hypothalamus. This part of our brain is concerned with emotions and instincts, feeding, fighting, fleeing, and sexual behavior. As MacLean observes, everything in this emotional system is either "agreeable or disagreeable". Survival depends on avoidance of pain and repetition of pleasure. The limbic system is the seat of the value judgments that we make, often unconsciously, that exert such a strong influence on our behavior.
- Lastly, the **neocortex**, or neomammalian brain, began its spectacular expansion in primates a few million years ago, as the genus *Homo* emerged. In modern humans, the neocortex takes up two thirds of the total brain mass. Although all

animals also have a neocortex, it is relatively small, with few or no convolutions (indicating less surface area, complexity, and development). A mouse without a cortex can act in fairly normal way (at least to a casual observer), whereas a human without a cortex is a vegetable. The neocortex is flexible and has almost infinite learning abilities. It has been responsible for the development of human language, abstract thought, imagination, and consciousness.

These three parts of the brain do not operate independently of one another. They have established numerous interconnections through which they influence one another. The neural pathways from the limbic system to the cortex, for example, are especially well developed. Thus, the human brain is the result of cumulative natural selection. Because older structures had proven their effectiveness for meeting certain fundamental needs in our evolutionary ancestors, there was no reason for them to disappear. Instead, evolution favoured a process of building expansions and additions, rather than rebuilding everything from the bottom up.

There are 12 pairs of cranial nerves, numbered I to XII starting with the most anterior. They also have descriptive names. Cranial nerves emerge from the brain and exit the cranium through its foramina. All except the vagus nerve (X), which serves abdominal viscera, travel to muscles and sense organs exclusively in the head and neck. Cranial nerves I and II are sensory and the rest are mixed nerves, although the only sensory functions of some of the mixed nerves are proprioception from the muscles they innervate. The cell bodies of sensory nerve fibers are located in ganglia outside the brain, while the cell bodies of motor nerve fibers are located in nuclei within the brain.[xvii]

- The olfactory nerve (I), a sensory nerve, is concerned with the sense of smell. It arises from the olfactory epithelium in the nasal cavities.
- The optic nerve (II), another sensory nerve, is concerned with the sense of vision. It arises from the retina.
- The oculomotor nerve (III) has both motor and proprioceptive fibers. It controls eye movements, opening of the eyelids, constriction of the pupil, and focusing. It arises from the midbrain.
- The trochlear nerve (IV) has both motor and proprioceptive fibers and is involved with eye movements. It also arises from the midbrain.
- The trigeminal nerve (V) has mostly sensory but also some motor and proprioceptive fibers. It is the largest of the cranial nerves and arises from the junction of the pons and medulla oblongata.
- The abducens nerve (VI) has both motor and proprioceptive fibers and is involved with eye movements. It, too, arises from the junction of the pons and the medulla.
- The facial nerve (VII) has motor, proprioceptive, and sensory fibers. It is the major motor nerve of facial expression and arises, again, from the junction of the pons and the medulla.
- The vestibulocochlear (formerly auditory) nerve (VIII) has mostly sensory with

---

[xvii] In the following description of individual cranial nerves, the term "sensory" refers to sensory functions other than proprioception.

some motor fibers. It is involved with audition (sense of hearing) and equilibrium (sense of balance). It arises primarily from the inner ear.

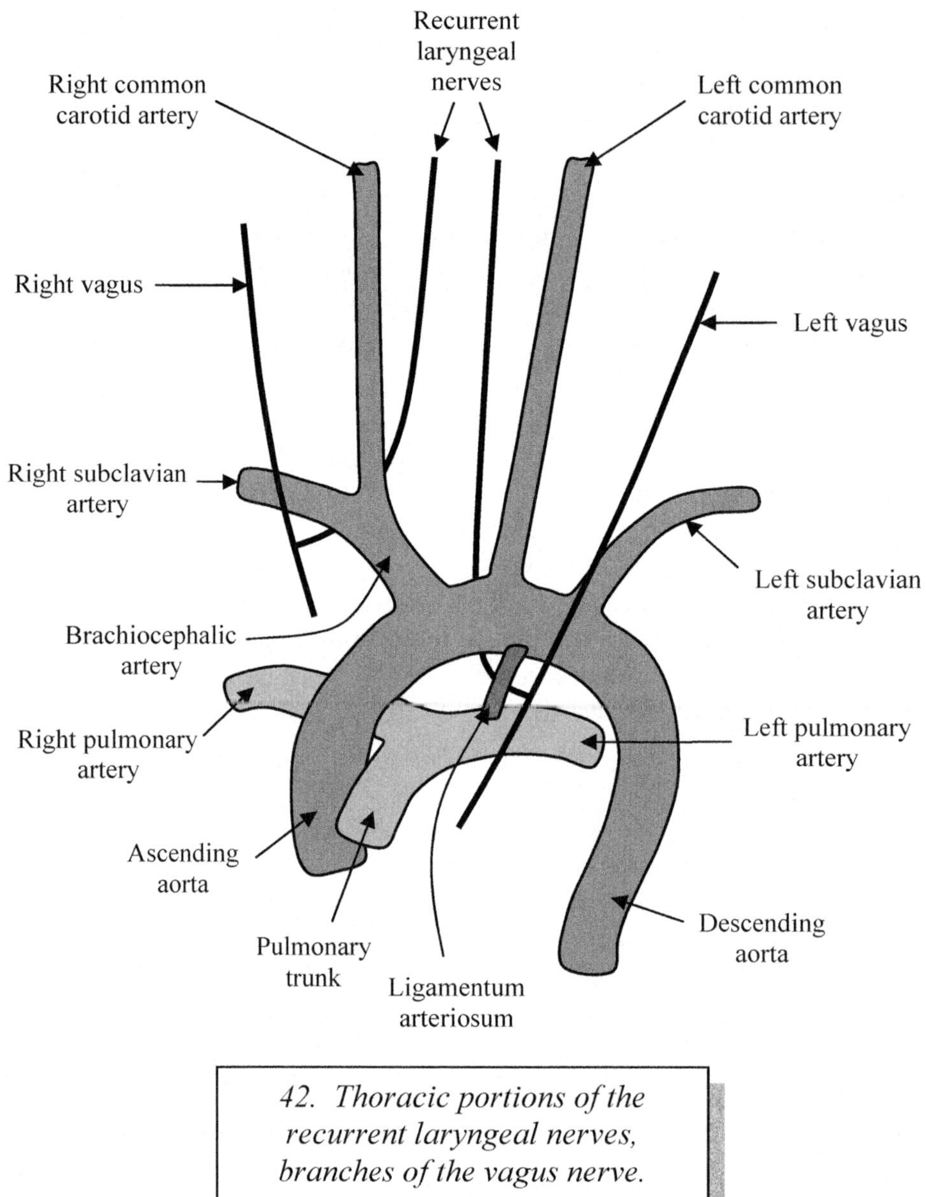

42. *Thoracic portions of the recurrent laryngeal nerves, branches of the vagus nerve.*

- The glossopharyngeal nerve (IX) has motor, proprioceptive, and sensory fibers. It is involved with diverse functions such as swallowing, regulation of blood pressure and respiration, and sensations of touch from the tongue, pharynx, and external ear. It arises from the medulla oblongata.
- The vagus nerve (X) has motor, proprioceptive, and sensory fibers. It is also involved with diverse functions, including swallowing, speech, and regulation of pulmonary, cardiovascular, and gastrointestinal activity. The vagus also arises

from the medulla, but is the only cranial nerve that extends beyond the head-neck region. It descends through the neck and thoracic region, sending branches to the tongue, carotid arteries, pharynx, larynx, and heart, and penetrates the diaphragm, sending branches to abdominal viscera.

- The accessory nerve (XI) has motor and proprioceptive fibers. It is involved with swallowing and movements of the head, neck, and shoulders. Cranial nerve XI, unlike all other cranial nerves, arises from **both** the brainstem (medulla oblongata) and the spinal cord.
- The hypoglossal nerve (XII) has predominantly motor with sparse proprioceptive fibers. It is involved with movements of the tongue for speech, food manipulation, and swallowing. It arises from the medulla.

One of the best known branches of the vagus nerve (X) is the recurrent laryngeal nerve (RLN), which supplies motor function and sensation to the larynx. The RLN arises after the vagus nerve descends through the neck, and then loops back up ("recurs") through the tracheo-esophageal groove to reach the larynx. The longer left RLN loops under and around the arch of the aorta before ascending. The shorter right RLN loops around the right subclavian artery before ascending. Damage to the RLN causes laryngeal palsy (paralysis of the larynx) on the affected side and can result from diseases inside the chest such as a tumor or an aneurysm of the arch of the aorta.

From an evolutionary perspective, the RLN in mammals is notable because it is an obvious case of poor design. As described above, it does not extend directly from cranium to larynx, making it much longer than necessary and more vulnerable to damage. For example, in a giraffe that means a 20-foot length of nerve is used where a one-foot length would have sufficed. In fish the homologous nerve follows the same path, but in fish the path is straight. It is easy to explain why the nerve takes this circuitous route in mammals if we accept that they evolved from fish-like ancestors.

TOPIC 24.5. THE PERIPHERAL NERVOUS SYSTEM

The PNS may be divided into three parts, or pathways: the somatic nervous system (SNS), the autonomic nervous system (ANS), and the enteric nervous system (ENS). The somatic nervous system produces motor responses that can be consciously controlled. It consists of sensory neurons that conduct nerve impulses from somatic and special sensory receptors (primarily in the head, body wall, and extremities) to the CNS and motor neurons that conduct nerve impulses from the CNS to skeletal muscles only.

The autonomic nervous system produces motor responses that are not normally under conscious control. Its sensory neurons conduct nerve impulses from autonomic sensory receptors, primarily in the viscera, to the CNS. Its motor neurons carry nerve impulses from the CNS to smooth muscle, cardiac muscle, and glands that are **not** located in the digestive tract. The ANS has two motor divisions, or pathways, that regulate the same organs: the sympathetic division is responsible for producing the "fight-or-flight" response and the parasympathetic division is responsible for recovering from the "fight-

or-flight" response.

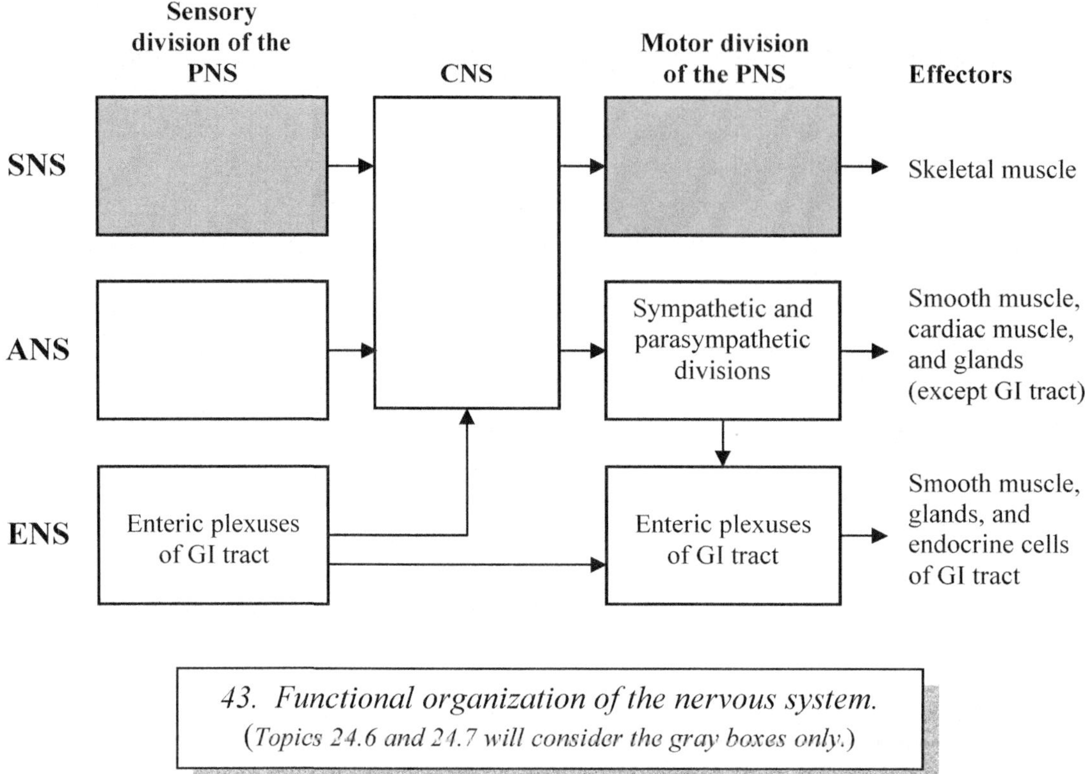

43. Functional organization of the nervous system.
(*Topics 24.6 and 24.7 will consider the gray boxes only.*)

The enteric nervous system, formerly considered part of the ANS and therefore involuntary in action, is the "brain of the gut". It is organized into enteric plexuses, which are networks of nerve fibers within the wall of the alimentary canal. Enteric plexuses include the submucosal (or Meissner's) plexus and the myenteric (or Auerbach's) plexus. Enteric sensory neurons conduct nerve impulses from sensory receptors in the digestive tract either to the CNS or to the enteric plexuses. Enteric sensory impulses to the plexuses result in motor impulses that travel to the smooth muscle, glands, and endocrine cells of the digestive tract. Enteric sensory impulses to the CNS result in motor impulses that first travel to the enteric plexuses and then to the digestive tract.

TOPIC 24.6. THE SOMATIC NERVOUS SYSTEM: GENERAL SENSES AND NEURAL PATHWAYS

A sensory receptor is any structure, such as sensory dendrites and sense organs, that is specialized to detect a stimulus. Sensation is the body's awareness of external or internal stimuli, while perception is conscious sensation. Perception always occurs in the cerebral cortex. Four events are usually necessary for a sensation to arise:

- Stimulation, in which a sufficiently strong stimulus activates specific sensory receptors.

- Transduction, in which a sensory receptor responds to the stimulus by producing a nerve impulse.
- Conduction, in which nerve impulses are carried to the CNS by sensory neurons.
- Translation, in which a region of the CNS converts the nerve impulses into a sensation.

Two important kinds of information transmitted by sensory receptors are stimulus modality (selectivity) and stimulus duration. Modality refers to the type of stimulus or sensation a sensory receptor responds to or produces, respectively. A receptor typically responds vigorously to one particular type of stimulus and weakly or not at all to others. Vision, audition, and gustation are examples of stimulus modalities. Duration is conveyed by the pattern of nerve impulses produced by a sensory receptor. All receptors other than pain receptors exhibit stimulus adaptation, which is a decrease in the frequency of nerve impulses and thus sensation during a prolonged stimulus. Adapting to hot bath water is an example.

The senses can be grouped into two major categories: general senses and special senses. The general senses have receptors that are widely distributed in the body rather than being limited to specific places. They can be divided into somatic senses and visceral senses. Somatic senses include touch, pressure, stretch, heat, cold, and pain, while visceral senses provide information about conditions within internal organs. The special senses, the receptors for which are localized in the body, are olfaction (sense of smell), gustation (sense of taste), vision (sense of sight), audition (sense of hearing), and equilibrium (sense of balance).

Sensory receptors may be classified in various, overlapping ways. One way is by their microscopic structures. Nonencapsulated (free) nerve endings are the bare dendrites of sensory neurons, such as the receptors for pain, tickle, and itch. Encapsulated nerve endings are sensory dendrites wrapped in glial cells or connective tissue, such as tactile (Meissner's) corpuscles or corpuscles of touch, muscle spindles, and tendon organs. Other receptors are separate cells that synapse with sensory neurons, such as hair cells and photoreceptors.

Receptors can also be classified based on their location. Exteroceptors are located at or near the body surface and provide information about the external environment. They include the receptors for vision, audition, gustation, olfaction, touch, and superficial somatic pain. Interoceptors are located in blood vessels, viscera, muscles, and the nervous system and provide information about the internal environment. They are responsible for feelings of visceral pain, nausea, stretch, and pressure. Proprioceptors are located in muscles, tendons, joints, and the inner ear and provide information about body position and movement. Proprioception, then, is the awareness of the position of the limbs in space, independent of vision. There are various types of proprioceptors. Muscle spindles, located in skeletal muscles, are encapsulated nerve endings that detect stretching of muscles. Tendon organs, located at junctions of tendon and muscle, are encapsulated nerve endings that detect excessive tension on tendons. Joint kinesthetic receptors, located within and around articular capsules of synovial joints, include lamellated corpuscles, Type II cutaneous mechanoreceptors (Ruffini corpuscles), tendon organs, and nonencapsulated nerve endings that detect joint movement.

Finally, receptors can be classified based on their stimulus modality. These include mechanoreceptors, thermoreceptors, photoreceptors, chemoreceptors, osmoreceptors, and nociceptors.

Mechanoreceptors respond to physical deformation of the plasma membrane caused by touch, pressure, stretch, tension, or vibration. The tactile corpuscles or corpuscles of touch, located in the dermal papillae of hairless skin, are encapsulated dendrites for fine touch. Hair root plexuses are nonencapsulated nerve endings wrapped around hair follicles and are stimulated by movements on the skin surface that disturb the hairs. Type I cutaneous mechanoreceptors (Merkel disks), abundant in the fingertips, hands, lips, and external genitalia, are nonencapsulated nerve endings for fine touch. Type II cutaneous mechanoreceptors, located deep in the dermis, are encapsulated receptors that detect stretching of the skin. Lamellated corpuscles, widely distributed in the body, are encapsulated dendrites that detect pressure, fast vibrations, and tickling. Other nonencapsulated nerve endings mediate the itch or tickle sensation.

Thermoreceptors are nonencapsulated nerve endings that respond to temperature and produce thermal sensation. Cold receptors, located in the stratum basale of the epidermis, detect temperatures between 10° and 40° C. Warm receptors, located in the dermis, detect temperatures between 32° and 48° C. Temperatures below 10° C and above 48° C stimulate mainly pain receptors.

Photoreceptors detect light that strikes the retina of the eye. Chemoreceptors detect chemicals in the oral cavity (gustation), nasal cavity (olfaction), and body fluids. Osmoreceptors sense the osmotic pressure of body fluids. Baroreceptors are stretch receptors located, for example, in the walls of blood vessels. By detecting the amount of stretch in the blood vessel walls, they determine the pressure of the blood flowing through the vessel and send this information to the central nervous system.

Nociceptors (pain receptors), which are located throughout the body (except the brain), are nonencapsulated nerve endings that respond to stimuli resulting from physical or chemical damage to tissue. They adapt slightly or not at all to a prolonged stimulus. There are two types of nociceptors that correspond to different pain sensations. Nociceptors with (myelinated) type A axons conduct nerve impulses rapidly and produce the sensation known as fast (acute) pain, such as needle puncture. Nociceptors with (nonmyelinated) type C axons conduct nerve impulses more slowly and produce the sensation known as slow (chronic) pain, such as a toothache. Slow pain gradually increases in intensity.

There are other types of pain in addition to fast and slow pain. Pain may also be classified according to where it originates as somatic or visceral. Somatic pain includes superficial somatic pain, which is due to stimulation of receptors in the skin, and deep somatic pain, which is due to stimulation of receptors in skeletal muscles, joints, tendons, and fascia. Visceral pain is due to stimulation of receptors in visceral organs. Referred pain is the term for pain that is felt in the skin overlying the stimulated organ or in a surface area far from the stimulated organ. Phantom limb sensation (phantom pain) is the term for pain that is felt in an amputated (nonexistent, phantom) limb.

Thousands of **somatic sensory pathways** conduct nerve impulses from somatic receptors below the head to the primary somatosensory area in the cerebral cortex or to the cerebellar cortex. They can be classified according to the ascending tracts they follow in the spinal cord and brainstem. Impulses to the cerebral cortex travel through two major groups of tracts: the **posterior column-medial lemniscus system** and the **spinothalamic system**. Impulses to the cerebellar cortex travel through the **spinocerebellar system**. Each pathway to the cerebral cortex consists of a series of three neurons called first-, second-, and third-order neurons. Each pathway to the cerebellar cortex consists of only two neurons, a first- and second-order.

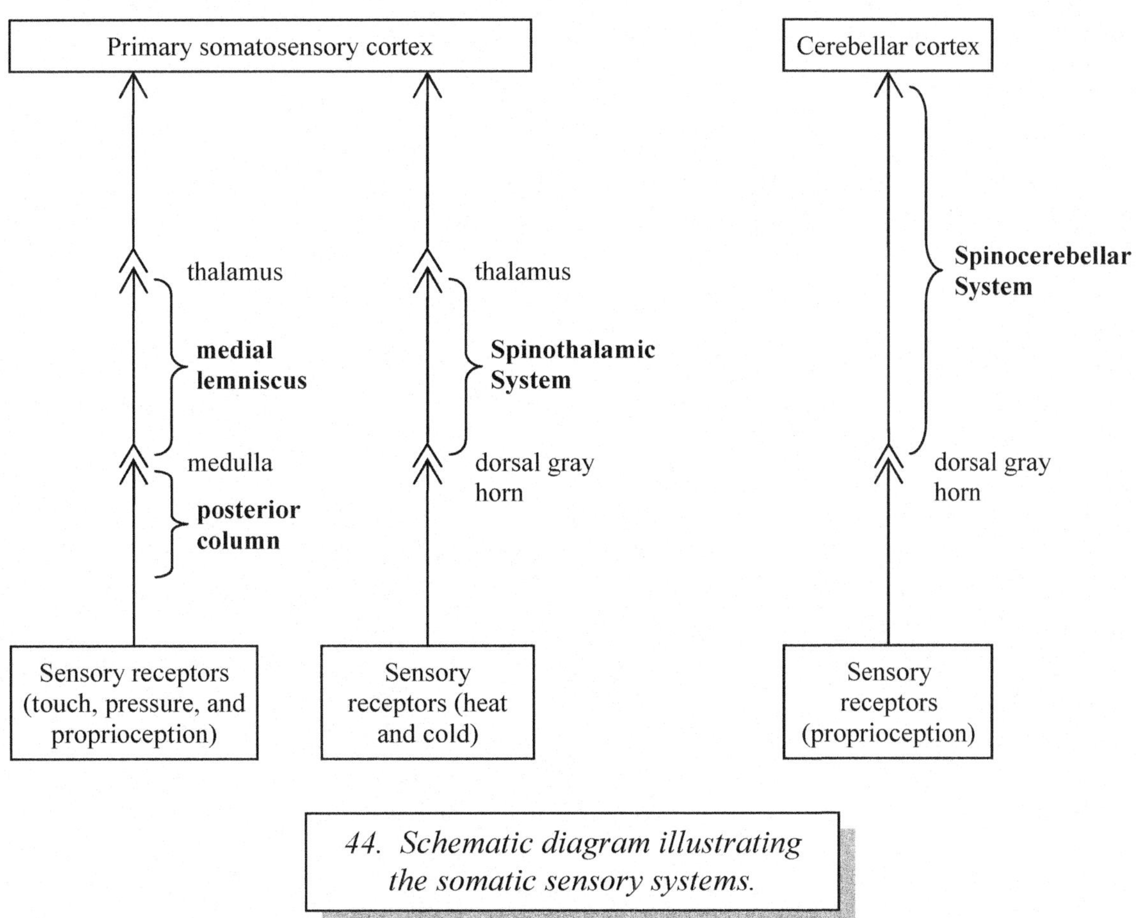

44. Schematic diagram illustrating the somatic sensory systems.

The posterior column-medial lemniscus system conducts impulses for touch, pressure, and proprioception. A somatic receptor generates impulses that travel to the spinal cord through a first-order neuron that enters the dorsal gray horn and ascends the posterior column to the medulla. Here it synapses with a second-order neuron. The axon of this neuron decussates and ascends through a tract called the medial lemniscus that terminates in the thalamus. Here this axon synapses with a third-order neuron whose axon extends to the primary somatosensory area of the cerebral cortex.

The spinothalamic system conducts impulses for heat and cold. A somatic receptor generates impulses that travel to the spinal cord by the axon of a first-order neuron, which synapses with a second-order neuron in the dorsal gray horn. The axon of this neuron decussates to the contralateral side of the spinal cord and ascends through either the **lateral** or **anterior spinothalamic tract** to the thalamus, where it synapses with a third-order neuron. The axon of this neuron extends to the primary somatosensory area of the cerebral cortex.

The spinocerebellar system conducts impulses from proprioceptors in the muscles and tendons of the legs and trunk to the cerebellar cortex, allowing the cerebellum to coordinate muscle action. The first-order neuron synapses in the dorsal gray horn with the second-order neuron, the axon of which then ascends through either the **posterior spinocerebellar tract** ipsilaterally or the **anterior spinocerebellar tract** contralaterally. The second-order neurons terminate in the cerebellum.

Thousands of **somatic motor pathways**, each consisting of an upper motor neuron (UMN) and a lower motor neuron (LMN), conduct nerve impulses from the brain to the skeletal muscles. The axons of UMNs extend from the brain to the LMNs (spinal nerves and cranial nerves) through one of two major groups of tracts: the **direct motor system** (or pyramidal system) and the **indirect motor system** (or extrapyramidal system).

The direct motor system contains axons of UMNs whose cell bodies all lie in the cerebral cortex, with 80 percent in the primary motor area. This system includes the corticospinal tracts and the corticobulbar tracts.

The corticospinal tracts conduct impulses from the cerebral cortex for the control of skeletal muscles below the head and neck. These tracts produce pyramid-shaped bulges on the anterior surface of the upper medulla oblongata; hence, they are also called the pyramidal tracts, although this term is becoming obsolete. About 85 to 90 percent of corticospinal axons decussate to the contralateral side of the lower medulla, forming the **lateral corticospinal tract**. The remaining 10 to 15 percent of corticospinal axons, which remain on the ipsilateral side, form the **anterior corticospinal tract**; however, they ultimately decussate at the spinal cord segment where they synapse with an interneuron or LMN.

The **corticobulbar tracts** in the midbrain conduct impulses from the cerebral cortex for the control of skeletal muscles in the head and neck. Not all axons in these tracts have decussated. Each axon synapses with an interneuron or LMN in the motor nuclei of nine pairs of cranial nerves in the pons and medulla oblongata.

The indirect motor system contains axons of UMNs whose cell bodies all lie in various nuclei in the brainstem; these UMNs in turn receive motor input from neurons in the basal nuclei, cerebellum, and cerebral cortex. This system includes the **tectospinal tract** and **rubrospinal tract**, which conduct nerve impulses for the control of contralateral skeletal muscles, and the **vestibulospinal tract** and **reticulospinal tract**, which conduct nerve impulses for the control of ipsilateral skeletal muscles.

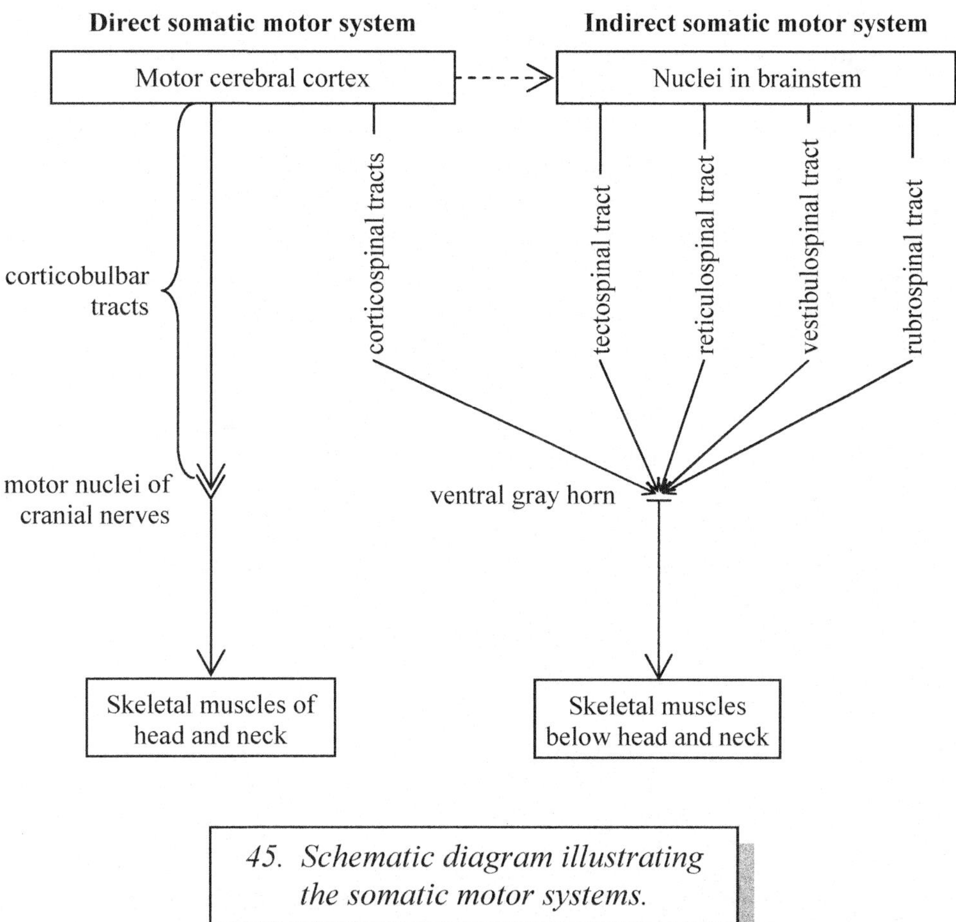

45. Schematic diagram illustrating the somatic motor systems.

Each LMN receives axons from descending tracts of both the direct and indirect systems. The synaptic connections of these axons with each LMN (often by way of interneurons) are in the thousands. The cell bodies of the LMNs are located either in the ventral gray horn of the spinal cord or in various motor nuclei of cranial nerves in the brainstem. Only LMNs conduct nerve impulses from the CNS to skeletal muscle fibers; hence, they are also known as the final common pathway.

TOPIC 24.7. THE SOMATIC NERVOUS SYSTEM: SPECIAL SENSES

Olfaction or olfactory sensation arises in the superior portion of the nasal cavity from the olfactory epithelium. The olfactory epithelium consists of three major types of cells: receptor, supporting, and basal cells. Olfactory receptor cells (chemoreceptors) are bipolar first-order sensory neurons that live for only one month. Cilia called olfactory hairs project from their dendrites and respond to odorants, inhaled chemicals with an odor that stimulate the receptors. The nonmyelinated axons of the receptors converge to form the right and left olfactory nerves. The olfactory nerves terminate in the olfactory bulbs of the brain. Supporting cells are simple columnar epithelial cells, while basal cells are stem cells that produce new olfactory receptor cells to replace those that die. Olfactory glands secrete mucus. The glands are located in connective tissue underlying the olfactory epithelium. The mucus moistens the surface of the olfactory epithelium and dissolves odorants.

Olfactory receptor (OR) genes provide the molecular basis for olfaction. About 60 percent of human OR genes are pseudogenes. In nonhuman apes, the fraction of OR pseudogenes is only approximately 30 percent. However, both the mouse and the dog have a significantly lower fraction of OR pseudogenes (about 20 percent) than do either humans or the other apes. Thus, there has been a decrease in the fraction of functional OR genes in apes relative to other mammals, with a further deterioration in humans. Consequently, human OR pseudogenes qualify as molecular vestigial structures.

Useless genes make no sense without an evolutionary explanation. The high fraction of OR pseudogenes in apes, including humans, may reflect a decreased reliance on the sense of smell in species for whom auditory and/or visual cues have become more important. For example, the dolphin, which evolved from land mammals, also carries many OR genes, but none are functional, since the dolphin no longer has any need to detect volatile odorants.[115] The progressive decrease in the fraction of functional OR genes in apes began about 23 million years ago, when they acquired trichromatic vision.[xviii, 116]

Gustation or gustatory sensation arises from taste buds located primarily in elevations on the tongue called papillae, with some taste buds also found in the soft palate, larynx, and pharynx. Each taste bud is an oval structure that consists of three types of cells: receptor, supporting, and basal cells. Gustatory receptor cells are chemoreceptors that synapse with first-order sensory neurons and live about 10 days. Supporting cells form a capsule that encloses the receptor cells, while basal cells are stem cells that replenish the gustatory receptor cells.

Vision or visual sensation arises from photoreceptors in the retina of the eye. The accessory structures of the eye include the: upper and lower eyelids (palpebrae); the eyelashes and eyebrows, which provide protection; and the lacrimal apparatus, a group of structures that produces and drains lacrimal fluid (tears). The lacrimal glands secrete lacrimal fluid, which washes over the surface of the eye and drains through the lacrimal canals into the lacrimal sac. From here, the fluid drains through the nasolacrimal duct into the nasal cavity. At the lateral angle of the eye, the ducts of the lacrimal gland open on its free surface. At the medial angle of the eye, there is a tiny membrane, called the semilunar fold or the plica semilunaris. As Darwin observed, "The nictitating membrane, or third eyelid, with its accessory muscles and other structures, is especially well developed in birds, and is of much functional importance to them, as it can be rapidly

---

[xviii] Humans, apes, and most, if not all, of the Old World monkeys have trichromatic vision, meaning that they can discriminate between the colors green, blue, and red.

drawn across the whole eyeball. It is found in some reptiles and amphibians, and in certain fishes, as in sharks. It is fairly well developed in the two lower divisions of the mammalian series, namely, in the [monotremes] and marsupials, and in some few of the higher mammals, as in the walrus. But in man, the Quadrumana[xix], and most other mammals, it exists, as is admitted by all anatomists, as a mere rudiment, called the semilunar fold."[117] Consequently, the plica semilunaris is a vestigial structure.

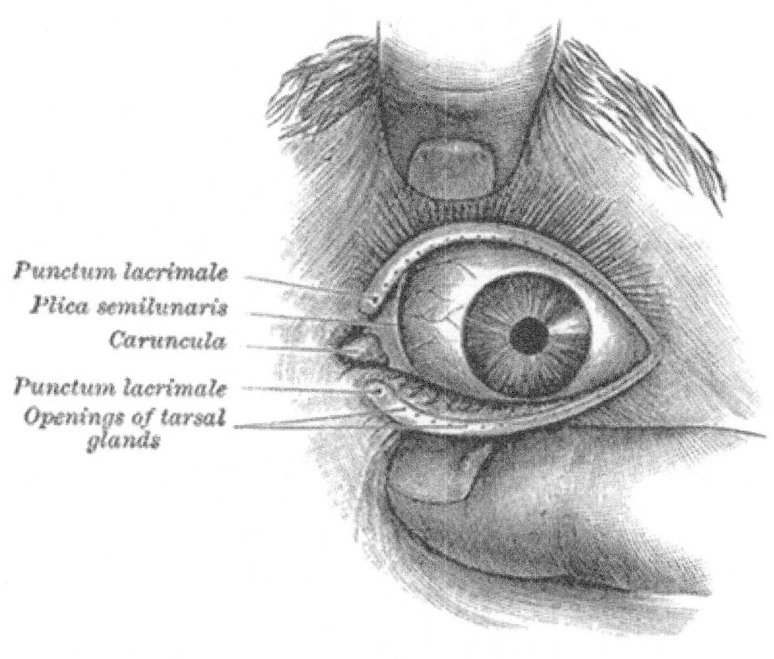

46. *Front of left eye with*
eyelids separated.
(Gray's Fig. 892)

The major parts of the eye are its wall and the lens. The wall of the eyeball consists of three layers:

- The outermost tunica fibrosa or fibrous tunic consists of the anterior cornea and the posterior sclera. The cornea, which refracts light, is a nonvascular, transparent, fibrous layer covering the colored iris. The sclera is a white layer of dense connective tissue that protects and supports the eyeball.
- The middle tunica vasculosa or vascular tunic (or uvea) consists of the choroid, the ciliary body, and the iris. The posterior choroid is the brown-black, highly vascular portion that lines most of the internal surface of the sclera and absorbs stray light. The ciliary body is a muscular ring of tissue between the choroid and

---

[xix]   At that time, the primates were divided into two orders: the Bimana, which included only humans, and the Quadrumana, which included all the other primates.

the iris. The anterior iris is the colored, circular structure composed of smooth muscle with a central hole called the pupil. By changing the diameter of the pupil, the iris varies the amount of light that enters the eye.
- The inner tunica nervosa consists of the smaller nonvisual retina, which is a non-photosensitive epithelium, and the larger optic retina. The optic retina lines the posterior three-quarters of the eyeball, while the nonvisual retina lines the ciliary body and the posterior surface of the iris. The boundary between the nonvisual retina and optic retina is a line called the ora serrata. At the center of the optic retina is the macula lutea. A small depression called the fovea centralis, the area of highest visual acuity (sharpness of vision), is located in the center of the macula lutea.

The lens is a nonvascular, transparent, oval structure located posterior to the iris. A ring of fibers called the suspensory ligament connects the lens to the ciliary body, which alters the shape of the lens in a process called visual accommodation. Accommodation fine-tunes focusing of light rays onto the retina. The lens divides the interior of the eyeball into two cavities:

- The smaller anterior cavity is located between the lens and cornea and is filled with aqueous humor, a watery fluid secreted by the ciliary body. The space between the cornea and iris is the anterior chamber, and the space between the iris and lens is the posterior chamber; they are continuous with each other via the pupil. Aqueous humor enters the posterior chamber and nourishes the lens. It then flows through the pupil into the anterior chamber, where it is reabsorbed. Glaucoma results when the rate of reabsorption is less than the rate of secretion.
- The larger posterior cavity called the vitreous chamber is filled with a gelatinous substance called the vitreous body. The vitreous body, unlike the aqueous humor, is not continuously replaced.

The retina forms from a cup-shaped outgrowth of the developing cerebrum in the embryo, so it is actually a part of the brain – the only part that can be viewed without dissection. The optic part of the retina consists of two major layers: the outer pigmented layer or pigment epithelium and the inner neural layer. The neural layer contains three layers of neurons:

- The inner ganglion cell layer borders the vitreous body. The axons of the ganglion cells converge toward the optic disc of the retina, where they unite to form the optic nerve that leaves the eye. Since the optic disc lacks the photoreceptor layer, it is the location of the blind spot, about 3 mm medial to the macula lutea. Bundled with the optic nerve are the central retinal artery and central retinal vein. The central retinal artery supplies capillaries on the anterior surface of the retina that nourish the ganglion and bipolar cell layers (see below).
- The middle bipolar cell layer contains three types of interneurons: bipolar cells, amacrine cells, and horizontal cells. They integrate sensory input from the photoreceptor cells before transmission to the ganglion cells.
- The outer photoreceptor cell layer consists of elaborately modified bipolar

neurons known as rods and cones. Rod cells are responsible for collecting dim intensities of light and are the main photoreceptors in night vision. Cone cells collect high intensity light and are responsible for producing color vision. Each photoreceptor cell has several parts, including an outer segment facing the pigment epithelium and a cell body closer to the bipolar cell layer. The outer segment is the photosensitive part.

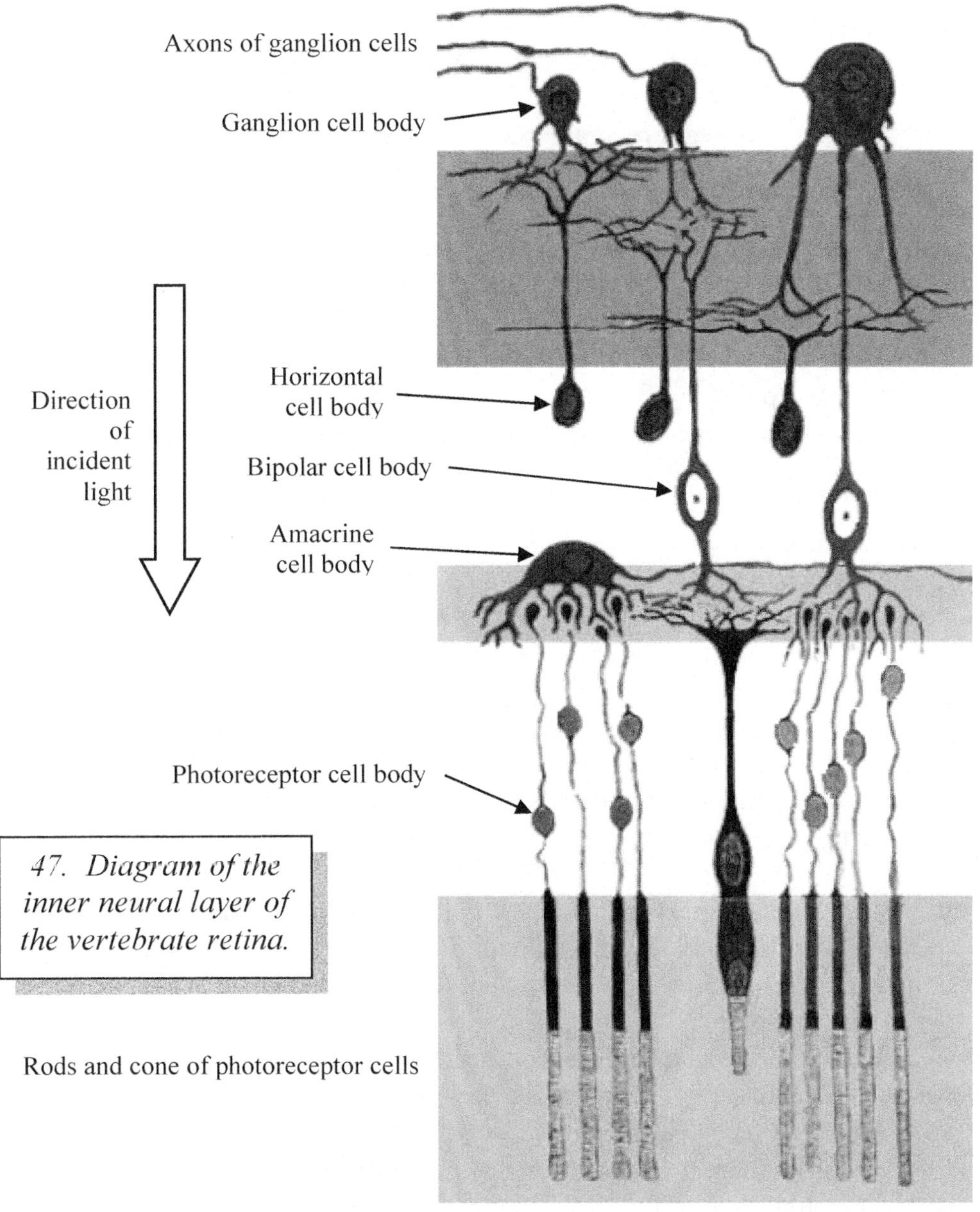

47. Diagram of the inner neural layer of the vertebrate retina.

Muller cells (not illustrated) are also found in the neural layer. They are analogous to the neuroglial cells of the CNS, providing structural support and perhaps nutrients for the neurons.

The cells of the pigment epithelium have microvilli that extend between the outer segments of the photoreceptor cells. These cells provide structural and metabolic support for the photoreceptor cells. They also absorb light rays passing through the retina, thus preventing back reflection. The choriocapillary layer of the tunica vasculosa nourishes the photoreceptor cells and the outer pigment epithelium.

The human eye, which is a typical vertebrate eye, is an example of poor design. The distinctive suboptimal features of the vertebrate eye include:

- The photoreceptors point toward the back of the (inverted) retina, away from incoming light, while neurons, axons, and blood vessels toward the front of the retina interfere with the passage of light.
- The fact that both photoreceptor cells and bipolar cells are nonmyelinated makes the retina more transparent, but also reduces the speed of conduction of nerve impulses.
- To improve visual acuity, the vertebrate eye focuses light on a patch of retina, the fovea centralis, where the blood vessels and axons have been pushed aside. However, the fovea is thus limited in size because of its reduced vascularization.
- The blind spot, where no image is formed, is created by the hole where the axons converge and penetrate through the retina to travel to the brain. Extra processing by the brain is required to "fill in" the missing part of the image, so that the blind spot is not normally apparent. More design to correct the initial design is surely not intelligent design!
- Because the axons of the optic nerve lie on the surface of the retina, the retina is loosely attached to the choroid. As a result, vertebrates have a high risk of retinal detachment, for example, due to a blow to the eye or head.
- The absorption of stray light by the pigment epithelium and the choroid generates heat next to the most thermally sensitive part of the photoreceptors, the outer segments. About 25 to 30 percent of the light falling on the retina ends up being absorbed by the choroid and re-radiated as heat. A **very** rapid blood flow through the choroidal circulation is therefore necessary to cool the choroid and prevent damage to the photoreceptors.[118, 119]

Like vertebrates, cephalopods (squids, cuttlefish, and octopi) and terrestrial gastropods (snails and slugs that live on land) have a single-lens eye structure with cornea and iris, similar to a camera. However, it has many advantages compared to the vertebrate eye:

- Cephalopods and terrestrial gastropods have their photoreceptors located at the surface of the (everted) retina, with their photosensitive outer segments facing the incoming light (see accompanying diagram[120]).
- Consequently, there are no neurons, axons, or blood vessels above the photoreceptors to interfere with the passage of light.
- The pigment layer is below the photoreceptors, in an area of dense blood vessels that can easily cool the tissue.[121] This means that the most thermally sensitive part of the retina is **not** next to the source of waste heat.
- There is no blind spot.

- Because there are no blood vessels and axons to push aside, an arbitrarily large "fovea" and greater visual acuity is possible. For example, cuttlefish have better visual acuity than cats.[122]
- The retina has far fewer parts than the retina of the vertebrate eye, because the axons of the photoreceptor cells form the optic nerve.
- Cephalopods have no risk of retinal detachment because the retina is anchored in place from below by axons of the optic nerve.[123]

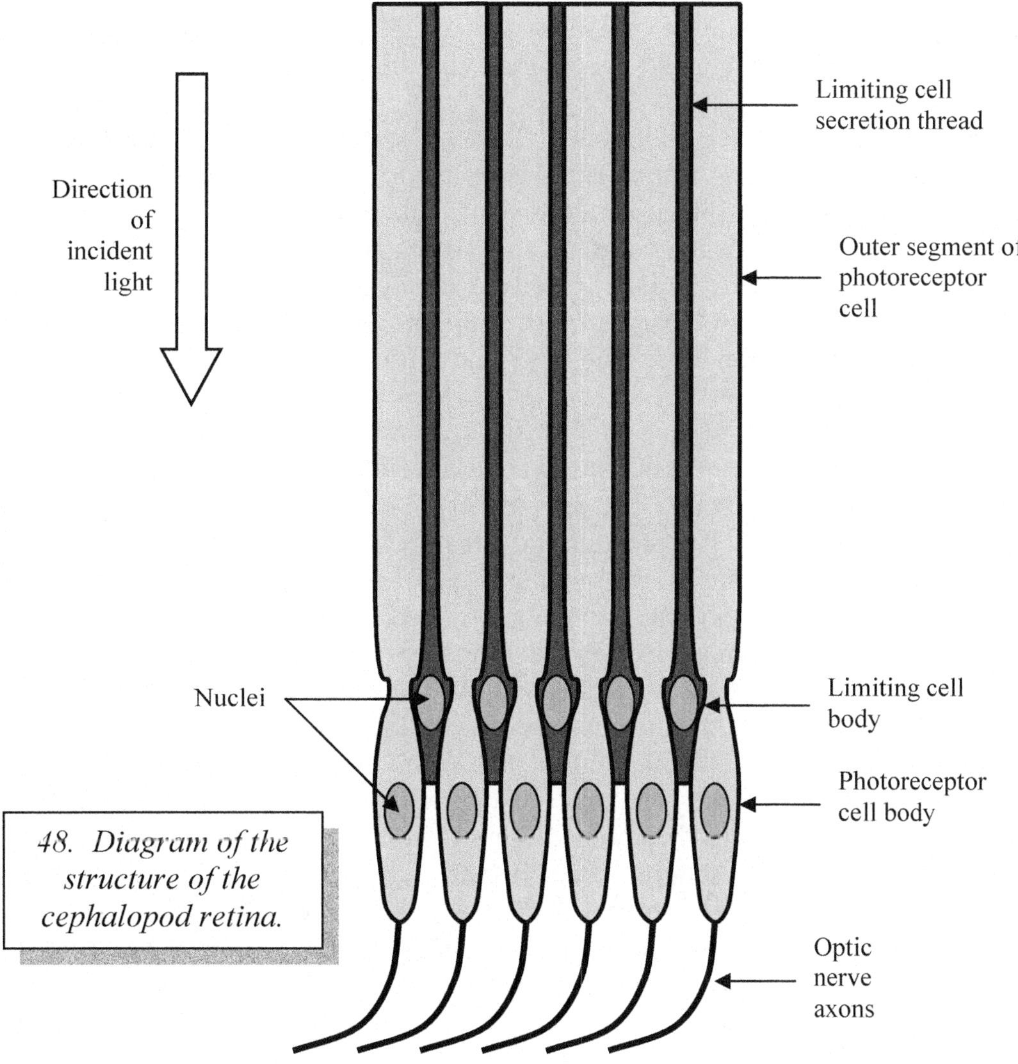

48. *Diagram of the structure of the cephalopod retina.*

Neither is the complex vertebrate eye better suited to terrestrial life, nor is the more efficient cephalopod eye better suited for murky underwater vision. This is supported by the fact that fish, though they are aquatic vertebrates with more species than all terrestrial vertebrates combined, have an inverted retina. Also, as noted above, terrestrial gastropods have an everted retina.

Evolution, which works by repeatedly modifying preexisting structures, can explain the inverted retina of vertebrates quite simply. The vertebrate retina evolved from brain cells that have wiring on the surface. Over time, evolution progressively modified this part of the brain for light sensitivity. Although the layer of light-sensitive cells gradually assumed a cup-like shape, it retained its original orientation, including a series of nerve connections on its surface. The cephalopod retina is better constructed because rather than evolving from brain cells, it evolved from skin cells, which retained their original orientation with the wiring below the surface. In this respect, the eyes of vertebrates and cephalopods are analogous, not homologous.

The ear is responsible for audition or auditory sensation (hearing) and equilibrium. It can be divided into three major regions: the external ear, the middle ear, and the internal ear.

The external (outer) ear, which collects sound waves and passes them inward, consists of the pinna, the external auditory canal, and the tympanic membrane. The pinna (auricle or auricula) is the flap-like, flexible structure on each side of the head. It consists of skin-covered elastic cartilage, except for the lowest portion, called the lobe or lobule. The external auditory canal, also called the external acoustic meatus, extends from the pinna to the thin tympanic membrane (eardrum); the outer portion is cartilaginous and the inner portion is bony. The eardrum closes the inner end of the auditory canal and separates it from the tympanic cavity.

The middle ear (tympanic cavity) is a small air-filled cavity, lined by mucosa, in the petrous part of the temporal bone. The opening to the external auditory canal is covered by the tympanic membrane. The most obvious feature of the middle ear are three tiny bones called the auditory ossicles: the malleus, incus, and stapes (so named for their resemblance to a hammer, anvil and stirrup, respectively). The malleus adheres to the tympanic membrane and is moved when the tympanic membrane is vibrated by sound waves. The malleus in turn causes the incus to move, and the incus transmits the vibrations to the stapes. The stapes touches a membrane covering the oval window, which is an opening to the fluid-filled inner ear. When the stapes moves, the fluid in the inner ear is vibrated, transmitting the sound waves to auditory receptors. Since fluid is incompressible, there must also be another membrane-covered outlet for the inner ear, allowing fluid to bulge out when the stapes vibrates. This opening, called the round window, is located in the wall of the tympanic cavity below the oval window. The membrane in the round window is called the secondary tympanic membrane. The auditory (pharyngotympanic or Eustachian) tube connects the middle ear to the nasopharynx and permits equalization of air pressure between the tympanic cavity and the surroundings. This prevents distortion of the tympanic membrane when the atmospheric air pressure changes.

The internal (inner) ear, also called the vestibulocochlear organ, is housed in a maze of passageways in the temporal bone called the bony labyrinth. The bony labyrinth can be divided into the following interconnected spaces: the vestibule, which is the central space next to the oval window; three semicircular canals, which extend posteriorly from the vestibule and each lie in a plane perpendicular to the other two; and the cochlea, a spiral canal that extends anteriorly from the vestibule. The bony labyrinth contains the membranous labyrinth (endolymphatic space), a system of interconnecting sacs and tubes filled with a fluid called endolymph. The space between the bony and membranous labyrinths is filled with perilymph, a fluid similar to cerebrospinal fluid.

The membranous labyrinth is composed of: two sacs called the utricle and saccule, which occupy the vestibule; three semicircular ducts, which occupy the semicircular canals; and the cochlear duct (scala media), which occupies the cochlea. The semicircular ducts, the utricle, and the saccule are collectively called the vestibular labyrinth, which is responsible for the sense of equilibrium. Six regions of sensory receptors project from the wall of the membranous labyrinth into the endolymph. Three cristae ampullaris, one in each semicircular duct, sense angular acceleration (turning) of the head. The macula of the utricle and the macula of the saccule sense the position of the head and linear acceleration. The spiral organ (organ of Corti) in the cochlear duct senses sound. All sensory receptor cells in the inner ear are nonneuronal mechanoreceptors called hair cells. Sound waves or movements of the head set the fluid within the inner ear in motion, which stimulates the hair cells.

The malleus and incus bones of the mammalian ear are homologous with much larger bones of the reptilian jaw.[124] In reptiles, each half of the mandible (lower jaw) consists of three bones: the dentary, the angular, and the articular. The articular bone of the mandible articulates with the quadrate bone of the skull, forming the quadro-articular joint. Most reptiles have a tympanic cavity, but it contains only the stapes bone, known as the columella auris.[125]

In mammals, each half of the mandible consists of just one bone, the dentary (they fuse together in humans). The dentary bone articulates with the squamosal bone of the skull, forming the dentary-squamosal joint. During embryonic development in all mammals, the articular of the lower jaw and the quadrate of the upper jaw become reduced in size and move into the ear to form the malleus and incus, respectively. In fact, the marsupial neonate, which is hardly more than an embryo but which needs functional jaws to fasten to the nipple in the pouch, has a reptilian jaw joint! The quadrate and articular bones move into the ear and take up the functions of the incus and malleus only **after** the dentary bone grows and contacts the squamosal bone of the skull, forming the mammalian jaw joint.

In the late nineteenth century, on the basis of the evidence from comparative anatomy and embryology, it was hypothesized that mammals evolved from reptiles. In 1912, the paleontologist Robert Broom predicted the nature of the necessary transitional forms in the evolution of the mammalian jaw joint from the reptilian condition. This prediction included an organism with a double jaw joint, one mammalian and the other reptilian. This was confirmed almost 50 years later by the discovery of fossils of *Diarthrognathus* and *Morganucodon* (mammal-like reptiles), which each have two functional jaw articulations, both the reptilian and the mammalian.[126, 127] Thus, the evolution of the human auditory ossicles is an example of cumulative natural selection.

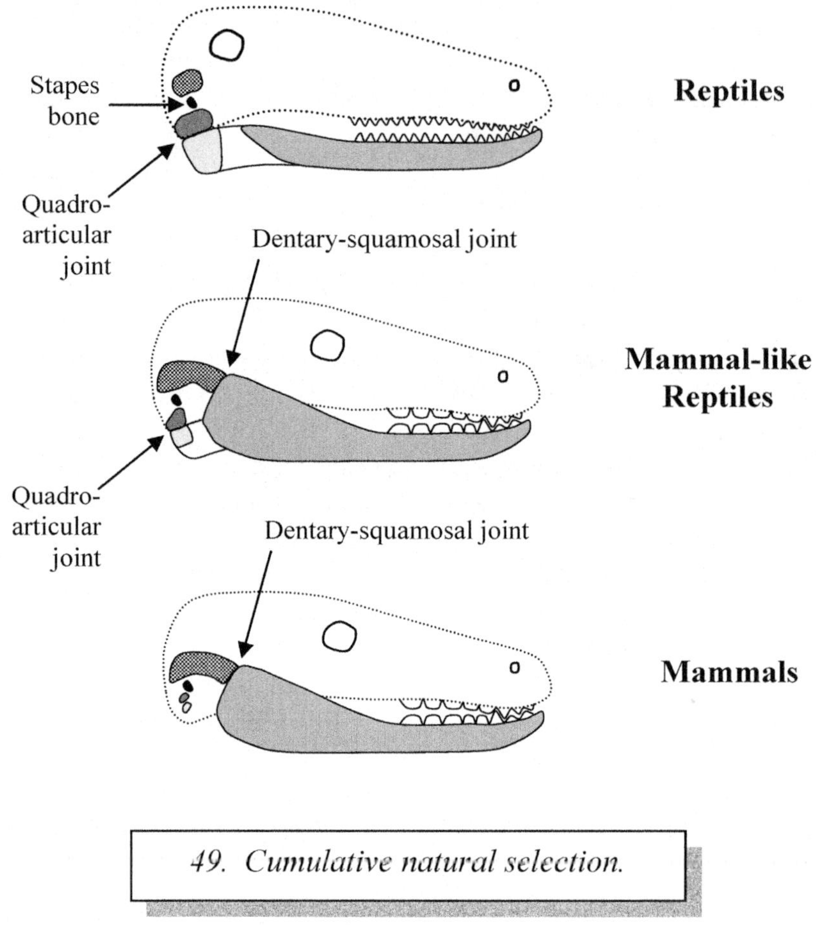

REFERENCES.

[1] Carl Sagan, *The Demon-Haunted World* (New York, NY: Random House, 1995) 309-17.

[2] Aristotle, *Metaphysics* IV 7: 1011b25-30.

[3] John A. Moore, *Science as a Way of Knowing* (Cambridge, MA: Harvard University Press, 1993) 136.

[4] Moore 94.

[5] Massimo Pigliucci, *Denying Evolution: Creationism, Scientism, and the Nature of Science* (Sunderland, MA: Sinauer Associates, Publishers, 2002) 138.

[6] Moti Ben-Ari, *Just a Theory: Exploring the Nature of Science* (Amherst, NY: Prometheus Books, 2005) 51-2.

[7] Bynum and Porter, 200.

[8] The National Academies of Science (NAS), *Science and Creationism: A View from the National Academies of Science*, 2/e (Washington, DC: The National Academies Press, 1999) 1-2. Also accessible online at <http://www.nap.edu>.

[9] NAS 1-2.

[10] Kenneth S. Saladin, *Anatomy and Physiology: The Unity of Form and Function* (New York, NY: McGraw-Hill Companies, 1998) 16.

[11] Charles Darwin, *The Origin of Species* 6/e (New York, NY: New American Library, 1958; originally published: London, England: John Murray, 1872) 404.

[12] NAS 1-2.

[13] The Society for the Study of Evolution, *Statement on Evolution by the Society for the Study of Evolution* (Aug 25, 2006). Accessed at <http://www.evolutionsociety.org/statements.htm> on May 22, 2007.

[14] Stephen Jay Gould, "Evolution as fact and theory," in *Science and Creationism* (New York, NY:

Oxford University Press, 1984) 118.

15 Sagan, *The Demon-Haunted World,* 304-306.

16 Bynum and Porter, 538.

17 Bynum and Porter, 177.

18 Sagan, *The Demon-Haunted World,* 210.

19 Sagan, *The Demon-Haunted World,* 298.

20 David Hume, *An Enquiry Concerning Human Understanding*: Section X (*Of Miracles*), Part I.

21 Carl Sagan, *Billions and Billions* (New York, NY: Ballantine Books, 1997) 60.

22 Bertrand Russell, *The Problems of Philosophy* (Mineola, NY: Dover Publications, Inc., 1999) 43.

23 Moore 405.

24 Sagan, *The Demon-Haunted World,* 212-16.

25 Moti Ben-Ari, *Just a Theory: Exploring the Nature of Science* (Amherst, NY: Prometheus Books, 2005) 174-8.

26 Jeffrey A. Lee, *The Scientific Endeavor* (San Francisco, CA: Addison Wesley Longman, 2000) 95-99.

27 Bynum and Porter, 201.

28 Pigliucci 129.

29 Lee 18.

30 Lewis Wolpert, *The Unnatural Nature of Science* (Cambridge, MA: Harvard Univ. Press, 1992) 95.

31 Bynum and Porter, 184.

32 M. Meselson and F. Stahl, "The replication of DNA in *E. coli,*" *Proceedings of the National Academy of Sciences USA* 44: 671-782 (1958).

33 Dennis O'Neil, *The primates: Humans* (November 21, 2007). Accessed at <http://anthro.palomar.edu/primate/prim_8.htm> on November 22, 2007.

34 For evolutionary relationships among these various hominid species, see the website of The Smithsonian Institution Human Origins Program at <http://www.mnh.si.edu/anthro/humanorigins/index2.htm>. Accessed October 2, 2007.

35 Ernst Mayr, *The Growth of Biological Thought: Diversity, Evolution, and Inheritance* (Cambridge, MA: Harvard University Press, 1982) 202.

36 Louis Agassiz, *Essay on Classification* (Mineola, NY: Dover Publications, Inc., 2004 [first published in 1857]) 187 n 33.

37 Moore 173.

38 Thomas Martin and Zhe-Xi Luo, "Homoplasy in the mammalian ear," *Science* 307(5711): 861-62 (Feb. 11, 2005).

39 Moore 177.

40 Agassiz 187 n 33.

41 Agassiz 12.

42 Mayr 426-7.

43 Mayr 510-525.

44 Charles Darwin, *The Variation of Animals and Plants under Domestication* vol. 2, 2/e (New York, NY: D. Appleton and Co., 1883) 206.

45 Martin Nickels, "Humans as a case study for the evidence of evolution," Dec. 31, 2001, Sept. 24, 2005 <www.talkorigins.org/faqs/homs/evidence_mn.html>.

46 Francois Jacob, "Evolution and tinkering" *Science* 196(4295): 1161-66 (June 10, 1977).

47 Gould, *Ever Since Darwin*, 91.

48 Stephen Jay Gould, *The Panda's Thumb: More Reflections in Natural History* (New York, NY: W.W. Norton & Co., 1980) 20-21.

49 Gould, *The Panda's Thumb*, 26.

50 Darwin, *The Origin of Species* 6/e, 423.

51 Darwin, *The Origin of Species* 6/e, 182.

52 Darwin, *The Origin of Species* 6/e, 419.

53 Charles Darwin, *The Descent of Man* 2/e (Amherst, New York: Prometheus Books, 1998; originally published: New York: Crowell, 1874) 161.

54 Elizabeth Culotta, "Chimp genome catalogs differences with humans," *Science* 309(5740): 1468-9 (Sept. 2, 2005).

[55] Jared Diamond, *The Third Chimpanzee: The Evolution and Future of the Human Animal* (New York, NY: HarperCollins Publishers, Inc., 1992) 23.

[56] Navin Elango, et al, "Variable molecular clocks in hominoids," *Proceedings of the National Academy of Sciences USA* 103(5): 1370-5 (Jan. 31, 2006).

[57] Bernard Wood and Brian G. Richmond, "Human evolution: Taxonomy and paleobiology," *Journal of Anatomy* 196: 19-60 (2000). Accessible at <www.gwu.edu/%7Ehebdp/BWPubs/120.pdf>.

[58] Tim Berra, *Evolution and the Myth of Creationism* (Stanford, CA: Stanford University Press, 1990) 99.

[59] Derek E. Wildman, et al, "Implications of natural selection in shaping 99.4% nonsynonymous DNA identity between humans and chimpanzees: Enlarging genus *Homo*," *Proceedings of the National Academy of Sciences USA* 100(12): 7181-8 (June 10, 2003).

[60] M. Gerstein and D. Zheng, "The real life of pseudogenes," *Scientific American* (Aug 2006) 49-55.

[61] M. Nishikimi et al, "Cloning and chromosomal mapping of the human nonfunctional gene for L-gulano-γ-lactone oxidase, the enzyme for L-ascorbic acid biosynthesis missing in man," *J. Biol. Chem.* 269: 13685-8 (1994).

[62] M. Nishikimi, et al, "Guinea pigs possess a highly mutated gene for L-gulono-γ-lactone oxidase, the key enzyme for L-ascorbic acid biosysthesis missing in this species," *J. Biol. Chem.* 267: 21967-72 (1992).

[63] Y. Ohta and M. Nishikimi, "Random nucleotide substitutions in primate nonfunctional gene for L-gulano-γ-lactone oxidase, the missing enzyme in L-ascorbic acid biosynthesis," *Biochimica et Biophysica Acta* 1472: 408-11 (1999).

[64] Edward E. Max, "Plagiarized Errors and Molecular Genetics," *The Talk.Origins Archive* May 5, 2003, April 7, 2006 <www.talkorigins.org/faqs/molgen/>.

[65] Massimo Pigliucci, 131.

[66] R. Perkins, et al, "Coccygectomy for severe refractory sacrococcygeal joint pain," *J Spinal Disord Tech* 16: 100-103 (2003).

[67] J. Selim, "Useless body parts: What do we need sinuses for, anyway?" *Discover* 25(6) (2004).

[68] G. McArdle, "Is inguinal hernia a defect in human evolution and would this insight improve concepts for methods of surgical repair?" *Clinical Anatomy* 10(1): 47-55 (1997).

[69] Diane M. Doran, "Comparative locomotor behavior of chimpanzees and bonobos: The influence of morphology on locomotion," *American Journal of Physical Anthropology* 91(1): 83-98 (1993).

[70] Accessed at <http://anthropology.si.edu/humanorigins/ha/al129.htm> on November 22, 2007.

[71] M.D. Leakey and R.L. Hay, "Pliocene footprints in the Laetoli beds at Laetoli, northern Tanzania," *Nature* 278: 317-323 (March 22, 1979).

[72] Stanford 87.

[73] Stanford 52.

[74] H.L. Shapiro, "Man – 500,000 years from now," *Natural History* November-December 1933, March 3, 2006 <www.naturalhistorymag.com/editors_pick/1933_11-12a_pick.html>.

[75] Craig Stanford, *Upright: The Evolutionary Key to Becoming Human* (New York, NY: Houghton Mifflin Co., 2003) 50.

[76] W.T. Fitch and D. Reby, "The descended larynx is not uniquely human," *Proc. Royal Soc. London B* 268: 1669-75 (2001).

[77] L.J. Boe, J.L. Heim, K. Honda, and S. Maeda, "The potential Neandertal vowel space was as large as that of modern humans," *Journal of Phonetics* 30: 465-484 (2002).

[78] Takeshi Nishimura, Akichika Mikami, Juri Suzuki, and Tetsuro Matsuzawa, "Descent of the larynx in chimpanzee infants," *PNAS* 100(12): 6930-6933 (June 10, 2003). Accessed 7 April 2009: <http://www.pnas.org/content/100/12/6930.full>.

[79] Charles Darwin, *The Origin of Species* 1/e (London, England: Penguin Books Ltd., 1985; originally published: London, England: John Murray, 1859) 221.

[80] Milton Hilderbrand, *Analysis of Vertebrate Structure*, 4/e (New York, NY: John Wiley & Sons, Inc., 1995) 244-5.

[81] P.M. O'Connor and L.P. Claessens, "Basic avian pulmonary design and flow-through ventilation in non-avian theropod dinosaurs," *Nature* 436: 253-6 (2005).

[82] Betsy Mason, "Birds get high on dinosaur breathing," *BioEd Online* Nov. 5, 2003, Feb. 24, 2006 <www.bioedonline.org/news/news.cfm?art=680>.

[83] Olshansky, et al, 51-55.

[84] A.R. Hargens, et al, "Gravitational haemodynamics and oedema prevention in the giraffe," *Nature*

329(6134): 59-60 (1987).

[85] Henry M. McHenry, "Tempo and mode in human evolution," *Proceedings of the National Academy of Sciences USA* 91: 6780-6 (July 1994).

[86] W. Zenker and S. Kubik, "Brain cooling in humans - anatomical considerations," *Anatomy and Embryology* 193: 1-13 (1996).

[87] D. Falk, "A good brain is hard to cool," *Natural History* 93(8): 65 (1993).

[88] Nicole Bergot, "The awful tooth: Humans keep evolving, but lose some wisdom," *Columbia News Service* April 5, 2004, Feb. 24, 2006 <www.jrn.columbia.edu/studentwork/cns/2004-04-05/609.asp>.

[89] Hansell H. Stedman, et al, "Myosin gene mutation correlates with anatomical changes in the human lineage," *Nature* 428: 415-8 (March 25, 2004).

[90] D.M. Hardin, Jr., 1999. "Acute appendicitis: review and update," *American Family Physician* 60(7): 2027-2034 (1999).

[91] Douglas Theobald, "The vestigiality of the human vermiform appendix: A modern reappraisal," *The Talk.Origins Archive* Feb. 3, 2005, Feb. 24, 2006 <www.talkorigins.org/faqs/vestiges/appendix.html>.

[92] Joseph McCabe, *The Story of Evolution* (Boston, MA: Small Maynard and Co., 1912) 264. Accessible at <www.gutenberg.org>.

[93] L. Werdelin and A. Nilsonne, "The evolution of the scrotum and testicular descent in mammals: A phylogenetic view," *Journal of Theoretical Biology* 196: 61-72 (1999).

[94] P.Z. Myers, "Descent of the testicle," *Pharyngula* June 20, 2004, Feb. 24, 2006 <pharyngula.org/index/weblog/comments/descent_of_the_testicle/>.

[95] Sentiel A. Rommel, et al, "Anatomical evidence for a countercurrent heat exchanger associated with dolphin testes," *The Anatomical Record* 232(1): 150-156 (1992).

[96] Nesse and Williams 100.

[97] Francois Jacob 1165.

[98] Robert Boyd and Joan B. Silk, *How Humans Evolved*, 3/e (New York, NY: W.W. Norton & Co., 2003) 467-8.

[99] Appendix A: Notes on Early Human Development. In *Monitoring Stem Cell Research* [World Wide Web site]. Bethesda, MD: National Institutes of Health, U.S. Department of Health and Human Services, 2004 [cited Sunday, February 10, 2008]. Available at < http://www.bioethics.gov/reports/stemcell/appendix_a.html>.

[100] Appendix A: Early Development. In *Stem Cell Information* [World Wide Web site]. Bethesda, MD: National Institutes of Health, U.S. Department of Health and Human Services, 2006 [cited Sunday, February 10, 2008]. Available at <http://stemcells.nih.gov/info/scireport/appendixa>.

[101] Charles K. Weichert, *Anatomy of the Chordates*, 3/e (New York, NY: McGraw-Hill Book Co., 1965) 211.

[102] Weichert 65.

[103] Darwin, *The Origin of Species* 6/e, 410, 416-417.

[104] Frank J. Sonleitner, "What's wrong with *Pandas*?" *NCSE Resource* Nov. 24, 2004, Feb. 24, 2006 <www.ncseweb.org/resources/articles/5423_59_sonleitner_what39s_wr_11_24_2004.asp>.

[105] H. Tyndale-Biscoe and M. Renfree, *Reproductive Physiology of Marsupials* (Cambridge, UK: Cambridge University Press, 1987).

[106] Wood and Richmond 22.

[107] Nesse and Williams 130, 200.

[108] Gould, *Ever Since Darwin*, 65.

[109] Gould, *Ever Since Darwin*, 72.

[110] Diamond 60.

[111] W.A.H. Rushton, "A theory of the effects of fibre size in medullated nerve," *J. Physiol.* 115: 101-122 (1951). Accessible at <http://www.pubmedcentral.nih.gov/articlerender.fcgi?artid=1392008>.

[112] Bruno Dubuc, "The brain," *The Brain from Top to Bottom*, accessed Nov. 16, 2007 < http://thebrain.mcgill.ca/flash/a/a_01/a_01_cr/a_01_cr_ana/a_01_cr_ana.html >.

[113] Dubuc, <http://thebrain.mcgill.ca/flash/d/d_05/d_05_cr/d_05_cr_her/d_05_cr_her.html>

[114] P.D. MacLean, "The triune brain, emotion, and scientific basis." In F.O. Schmitt (ed.), *The Neurosciences: Second Study Programme* (New York: Rockefeller University Press, 1970).

[115] J. Freitag, et al, "Olfactory receptors in aquatic and terrestrial vertebrates," *Journal of Comparative Physiology* 183: 635-50 (1998).

[116] Yoav Gilad, et al, "Loss of olfactory receptor genes coincides with the acquisition of full trichromatic vision in primates," *Public Library of Science, Biology* 2(1):e5 (Jan. 20, 2004). Accessible at <biology.plosjournals.org>.

[117] Darwin, *The Descent of Man* 2/e, 17.

[118] L.M. Parver, et al, "The stabilizing effect of the choroidal circulation on the temperature environment of the macula," *Retina* 2(2): 117-20 (1982).

[119] L.M. Parver, "Temperature modulating action of choroidal blood flow," *Eye* 5(2): 181-5 (1991).

[120] D.H. Tompsett, *Sepia. Liverpool Marine Biological Committee Memoirs, 32* (London: Williams and Norgate, 1939), adapted from figure 71.

[121] S. Matsui, et al, "Adaptation of a deep-sea cephalopod to the photic environment: Evidence for three visual pigments," *J. Gen. Physiol.* 92(1): 55-66 (1988).

[122] F. Schaeffel, et al, "Accommodation in the cuttlefish (*Sepia officinalis*)," *J. Exp. Biol.* 202(22): 3127-34 (1999).

[123] Nesse and Williams 129.

[124] Thomas Martin and Zhe-Xi Luo, "Homoplasy in the mammalian ear," *Science* 307(5711): 861-2 (Feb. 11, 2005).

[125] Weichert 669-72.

[126] Gould, *Eight Little Piggies*, 95-108.

[127] Moore 176-177, 412-414.

ON THE RELATIONS OF MAN TO THE LOWER ANIMALS
by Thomas Henry Huxley[xx]

Multis videri poterit, majorem esso differentiam Simiæ et Hominis, quam diei et noctis; verum tamen hi, comparatione instituta inter summos Europæ Heroës et Hottentottes ad Caput bonæ spei degentes, difficillime sibi persuadebunt, has eosdem habere natales; vel si virginem nobilem aulicam, maxime comtam et humanissimam, conferre vellent cum homine sylvestri et sibi relicto, vix augurari possent, hunc et illam ejusdem esse speciei.–Linnæi Amoœnitates Acad. "Anthropomorpha."

The question of questions for mankind–the problem which underlies all others, and is more deeply interesting than any other–is the ascertainment of the place which Man occupies in nature and of his relations to the universe of things. Whence our race has come; what are the limits of our power over nature, and of nature's power over us; to what goal we are tending; are the problems which present themselves anew and with undiminished interest to [78][xxi] every man born into the world. Most of us, shrinking from the difficulties and dangers which beset the seeker after original answers to these riddles, are contented to ignore them altogether, or to smother the investigating spirit under the feather-bed of respected and respectable tradition. But, in every age, one or two restless spirits, blessed with that constructive genius, which can only build on a secure foundation, or cursed with the spirit of mere scepticism, are unable to follow in the well-worn and comfortable track of their forefathers and contemporaries, and unmindful of thorns and stumbling-blocks, strike out into paths of their own. The sceptics end in the infidelity which asserts the problem to be insoluble, or in the atheism which denies the existence of any orderly progress and governance of things: the men of genius propound solutions which grow into systems of Theology or of Philosophy, or veiled in musical language which suggests more than it asserts, take the shape of the Poetry of an epoch.

Each such answer to the great question, invariably asserted by the followers of its propounder, if not by himself, to be complete and final, remains in high authority and esteem, it may be for one century, or it may be for twenty: but, as invariably, Time proves each reply to have been a mere approximation to the truth–tolerable chiefly on account of the ignorance of those by whom it was accepted, and wholly intolerable [79] when tested by the larger knowledge of their successors.

In a well-worn metaphor, a parallel is drawn between the life of man and the metamorphosis of the caterpillar into the butterfly; but the comparison may be more just as well as more novel, if for its former term we take the mental progress of the race. History shows that the human mind, fed by constant accessions of knowledge, periodically grows too large for its theoretical coverings, and bursts them asunder to appear in new habiliments, as the feeding and growing grub, at intervals, casts its too narrow skin and assumes another, itself but temporary. Truly the imago state of Man seems to be terribly distant, but every moult is a step gained, and of such there have been many.

[xx]   T.H. Huxley, *Collected Essays*, vol. VII, *Man's Place in Nature* (London, England: D. Appleton & Co., 1893-4) 77-156. This essay was based on lectures that Huxley delivered in 1860.
[xxi]   Bracketed numbers refer to pagination in Huxley's published essay.

Since the revival of learning, whereby the Western races of Europe were enabled to enter upon that progress towards true knowledge, which was commenced by the philosophers of Greece, but was almost arrested in subsequent long ages of intellectual stagnation, or, at most, gyration, the human larva has been feeding vigorously, and moulting in proportion. A skin of some dimension was cast in the 16th century, and another towards the end of the 18th, while, within the last fifty years, the extraordinary growth of every department of physical science has spread among us mental food of so nutritious and stimulating a [80] character that a new ecdysis seems imminent. But this is a process not unusually accompanied by many throes and some sickness and debility, or, it may be, by graver disturbances; so that every good citizen must feel bound to facilitate the process, and even if he have nothing but a scalpel to work withal, to ease the cracking integument to the best of his ability.

In this duty lies my excuse for the publication of these essays. For it will be admitted that some knowledge of man's position in the animate world is an indispensable preliminary to the proper understanding of his relations to the universe; and this again resolves itself, in the long run, into an inquiry into the nature and the closeness of the ties which connect him with those singular creatures whose history has been sketched in the preceding pages.

The importance of such an inquiry is indeed intuitively manifest. Brought face to face with these blurred copies of himself, the least thoughtful of men is conscious of a certain shock, due perhaps, not so much to disgust at the aspect of what looks like an insulting caricature, as to the awakening of a sudden and profound mistrust of time-honoured theories and strongly-rooted prejudices regarding his own position in nature, and [81] his relations to the under-world of life; while that which remains a dim suspicion for the unthinking, becomes a vast argument, fraught with the deepest consequences, for all who are acquainted with the recent progress of the anatomical and physiological sciences.

I now propose briefly to unfold that argument, and to set forth, in a form intelligible to those who possess no special acquaintance with anatomical science, the chief facts upon which all conclusions respecting the nature and the extent of the bonds which connect man with the brute world must be based: I shall then indicate the one immediate conclusion which, in my judgment, is justified by those facts, and I shall finally discuss the bearing of that conclusion upon the hypotheses which have been entertained respecting the Origin of Man.

The facts to which I would first direct the reader's attention, though ignored by many of the professed instructors of the public mind, are easy of demonstration and are universally agreed to by men of science; while their significance is so great, that whoso has duly pondered over them will, I think, find little to startle him in the other revelations of Biology. I refer to those facts which have been made known by the study of Development.

It is a truth of very wide, if not of universal, [82] application, that every living creature commences its existence under a form different from, and simpler than, that which it eventually attains.

The oak is a more complex thing than the little rudimentary plant contained in the acorn; the caterpillar is more complex than the egg; the butterfly than the caterpillar; and each of these beings, in passing from its rudimentary to its perfect condition, runs through a series of changes, the sum of which is called its Development. In the higher animals these changes are extremely complicated; but, within the last half century, the labours of such men as Von Baer, Rathke, Reichert, Bischoff, and Remak, have almost completely unravelled them, so that the successive stages of development which are exhibited by a Dog, for example, are now as well known to the embryologist as are the steps of the metamorphosis of the silk-worm moth to the school-boy. It will be useful to consider with attention the nature and the order of the stages of canine development, as an example of the process in the higher animals generally.

The dog, like all animals, save the very lowest (and further inquiries may not improbably remove the apparent exception), commences its existence as an egg: as a body which is, in every sense, as much an egg as that of a hen, but is devoid of that accumulation of nutritive matter which confers upon the bird's egg its exceptional size and [83] domestic utility; and wants the shell, which would not only be useless to an animal incubated within the body of its parent, but would cut it off from access to the source of that nutriment which the young creature requires, but which the minute egg of the mammal does not contain within itself.

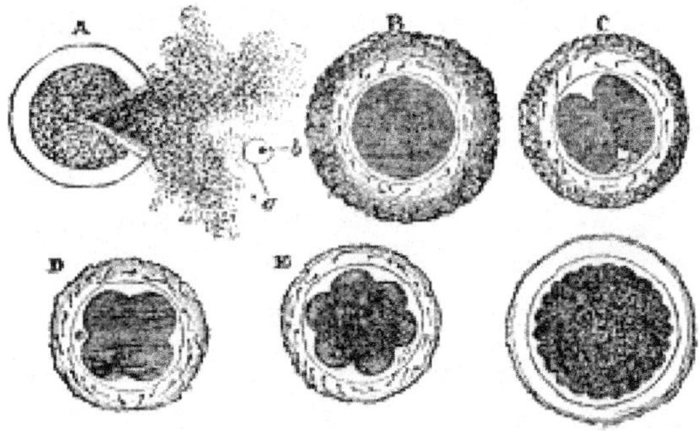

Fig. 13.–A. Egg of the Dog, with the vitelline membrane burst, so as to give exit to the yelk, the germinal vesicle (*a*), and its included spot (*b*). B. C. D. E. F. Successive changes of the yelk indicated in the text. After Bischoff.

The Dog's egg is, in fact, a little spheroidal bag (Fig. 13), formed of a delicate transparent membrane called the *vitelline membrane,* and about 1/130th to 1/120th of an inch in diameter. It contains a [84] mass of viscid nutritive matter–the *yelk*–within which is enclosed a second much more delicate spheroidal bag, called the *germinal vesicle (a).* In this, lastly, lies a more solid rounded body, termed the *germinal spot (b).*

The egg, or Ovum is originally formed within a gland, from which, in due season, it becomes detached, and passes into the living chamber fitted for its protection and maintenance during the protracted process of gestation. Here, when subjected to the required conditions, this minute and apparently insignificant particle of living matter becomes animated by a new and mysterious activity. The germinal vesicle and spot cease to be discernible (their precise fate being one of the yet unsolved problems of embryology), but the yelk becomes circumferentially indented, as if an invisible knife had been drawn round it, and thus appears divided into two hemispheres (Fig. 13, C).

By the repetition of this process in various planes, these hemispheres become subdivided, so that four segments are produced (D); and these, in like manner, divide and subdivide again, until the whole yelk is converted into a mass of granules, each of which consists of a minute spheroid of yelk-substance, inclosing a central particle, the so-called *nucleus* (F). Nature, by this process, has attained much the same result as that which a human artificer arrives at by his [85] operations in a brick-field. She takes the rough plastic material of the yelk and breaks it up into well-shaped tolerably even-sized masses–handy for building up into any part of the living edifice.

Next, the mass of organic bricks, or *cells* as they are technically called, thus formed, acquires an orderly arrangement, becoming converted into a hollow spheroid with double walls. Then, upon one side of this spheroid, appears a thickening, and, by and bye, in the centre of the area of thickening, a straight shallow groove (Fig. 14, A) marks the central line of the edifice which is to be raised, or, in other words, indicates the position of the middle line of the body of the future dog. The substance bounding the groove on each side next rises up into a fold, the rudiment of the side wall of that long cavity, which will eventually lodge the spinal marrow and the brain; and in the floor of this chamber appears a solid cellular cord, the so-called *notochord.* One end of the enclosed cavity dilates to form the head (Fig. 14, B), the other remains narrow, and eventually becomes the tail; the side walls of the body are fashioned out of the downward continuation of the walls of the groove; and from them, by and bye, grow out little buds which, by degrees, assume the shape of limbs. Watching the fashioning process stage by stage, one is forcibly reminded of the modeller in clay. Every part, every organ, is at first, as it were [86] pinched up rudely, and sketched out in the rough; then shaped more accurately; and only, at last, receives the touches which stamp its final character.

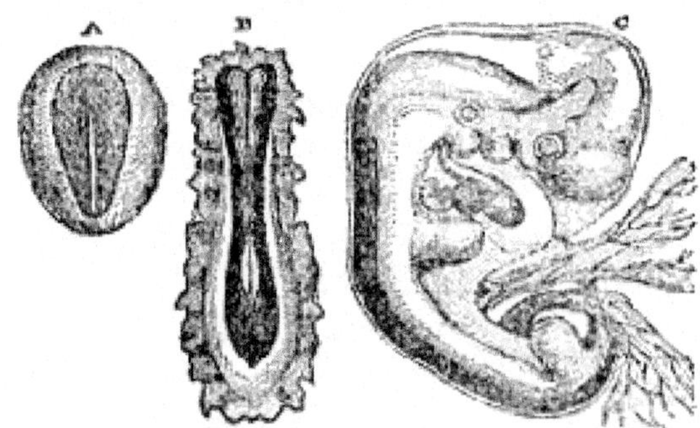

Fig. 14.–Earliest rudiment of the Dog. B. Rudiment further advanced, showing the foundations of the head, tail, and vertebral column. C. The very young puppy, with attached ends of the yelk-sac and allantois, and invested in the amnion.

Thus, at length, the young puppy assumes such a form as is shown in Fig. 14, C. In this condition it has a disproportionately large head, as dissimilar to that of a dog as the bud-like limbs are unlike his legs.

[87] The remains of the yelk, which have not yet been applied to the nutrition and growth of the young animal, are contained in a sac attached to the rudimentary intestine, and termed the yelk sac, or *umbilical vesicle.* Two membranous bags, intended to subserve respectively the protection and nutrition of the young creature, have been developed from the skin and from the under and hinder surface of the body; the former, the so-called *amnion,* is a sac filled with fluid, which invests the whole body of the embryo, and plays the part of a sort of water-bed for it; the other, termed the *allantois,* grows out, loaded with blood-vessels, from the ventral region, and eventually applying itself to the walls of the cavity, in which the developing organism is contained, enables these vessels to become the channel by which the stream of nutriment, required to supply the wants of the offspring, is furnished to it by the parent.

The structure which is developed by the interlacement of the vessels of the offspring with those of the parent, and by means of which the former is enabled to receive nourishment and to get rid of effete matters, is termed the *Placenta.*

It would be tedious, and it is unnecessary for my present purpose, to trace the process of development further; suffice it to say, that, by a long and gradual series of changes, the rudiment here depicted and described, becomes a puppy, is [88] born, and then, by still slower and less perceptible steps, passes into the adult Dog.

There is not much apparent resemblance between a barn-door Fowl and the Dog who protects the farm-yard. Nevertheless the student of development finds, not only that the chick commences its existence as an egg, primarily identical, in all essential respects, with that of the Dog, but that the yelk of this egg undergoes division–that the primitive groove arises, and that the contiguous parts of the germ are fashioned, by precisely similar methods, into a young chick, which, at one stage of its existence, is so like the nascent Dog, that ordinary inspection would hardly distinguish the two.

The history of the development of any other vertebrate animal, Lizard, Snake, Frog, or Fish, tells the same story. There is always, to begin with, an egg having the same essential structure as that of the Dog:–the yelk of that egg always undergoes division, or *segmentation* as it is often called: the ultimate products of that segmentation constitute the building materials for the body of the young animal; and this is built up round a primitive groove, in the floor of which a notochord is developed. Furthermore, there is a period in which the young of all these animals resemble one another, not merely in outward form, but in all essentials of structure, so closely, that the [89] differences between them are inconsiderable, while, in their subsequent course they diverge more and more widely from one another. And it is a general law, that, the more closely any animals resemble one another in adult structure, the longer and the more intimately do their embryos resemble one another: so that, for example, the embryos of a Snake and of a Lizard remain like one another longer than do those of a Snake and of a Bird; and the embryo of a Dog and of a Cat remain like one another for a far longer period than do those of a Dog and a Bird; or of a Dog and an Opossum; or even than those of a Dog and a Monkey.

Thus the study of development affords a clear test of closeness of structural affinity, and one turns with impatience to inquire what results are yielded by the study of the development of Man. Is he something apart? Does he originate in a totally different way from Dog, Bird, Frog, and Fish, thus justifying those who assert him to have no place in nature and no real affinity with the lower world of animal life? Or does he originate in a similar germ, pass through the same slow and gradually progressive modifications, depend on the same contrivances for protection and nutrition, and finally enter the world by the help of the same mechanism? The reply is not doubtful for a moment, and has not been doubtful any time these thirty years. Without question, the mode of origin and the early stages of the development of man are [90] identical with those of the animals immediately below him in the scale:–without a doubt, in these respects, he is far nearer the Apes, than the Apes are to the Dog.

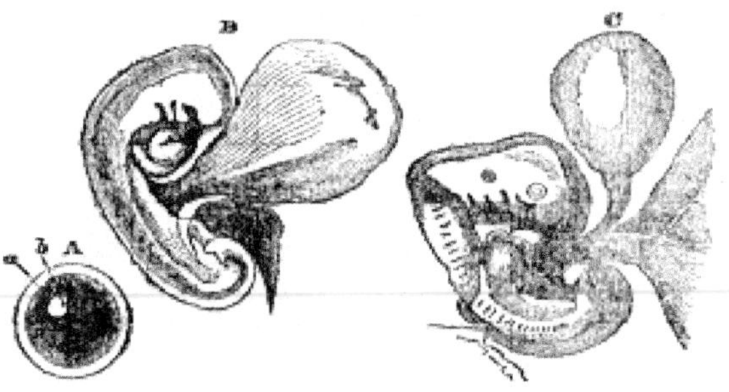

Fig. 15.–A. Human ovum (after Kölliker). *a..* germinal vesicle. *b.* germinal spot. B. A very early condition of Man, with yelk-sac, allantois and amnion (original). C. A more advanced stage (after Kölliker), compare Fig. 14, C.

The Human ovum is about l/125th of an inch in diameter, and might be described in the same terms as that of the Dog, so that I need only refer to the figure illustrative (15 A) of its structure. It leaves the organ in which it is formed in a similar fashion and enters the organic chamber prepared for its reception in the same way, the conditions of its development being in all respects the same. It has not yet been possible (and only [91] by some rare chance can it ever be possible) to study the human ovum in so early a developmental stage as that of yelk division, but there is every reason to conclude that the changes it undergoes are identical with those exhibited by the ova of other vertebrated animals; for the formative materials of which the rudimentary human body is composed, in the earliest conditions in which it has been observed, are the same as those of other animals. Some of these earliest stages are figured below and, as will be seen, they are strictly comparable to the very early states of the Dog; the marvellous correspondence between the two which is kept up, even for some time, as development advances, becoming apparent by the simple comparison of the figures with those on page 86.

Indeed, it is very long before the body of the young human being can be readily discriminated from that of the young puppy; but, at a tolerably early period, the two become distinguishable by the different form of their adjuncts, the yelk-sac and the allantois. The former, in the Dog, becomes long and spindle-shaped, while in Man it remains spherical: the latter, in the Dog, attains an extremely large size, and the vascular processes which are developed from it and eventually give rise to the formation of the placenta (taking root, as it were, in the parental organism, so as to draw nourishment therefrom, as the root of a tree extracts it from the soil) are arranged in an en[92]circling zone, while in Man, the allantois remains comparatively small, and its vascular rootlets are eventually restricted to one disk-like spot. Hence, while the placenta of the Dog is like a girdle, that of Man has the cake-like form, indicated by the name of the organ.

But, exactly in those respects in which the developing Man differs from the Dog, he resembles the ape, which, like man, has a spheroidal yelk-sac and a discoidal, sometimes partially lobed, placenta. So that it is only quite in the later stages of development that the young human being presents marked differences from the young ape, while the latter departs as much from the dog in its development, as the man does.

Startling as the last assertion may appear to be, it is demonstrably true, and it alone appears to me sufficient to place beyond all doubt the structural unity of man with the rest of the animal world, and more particularly and closely with the apes…

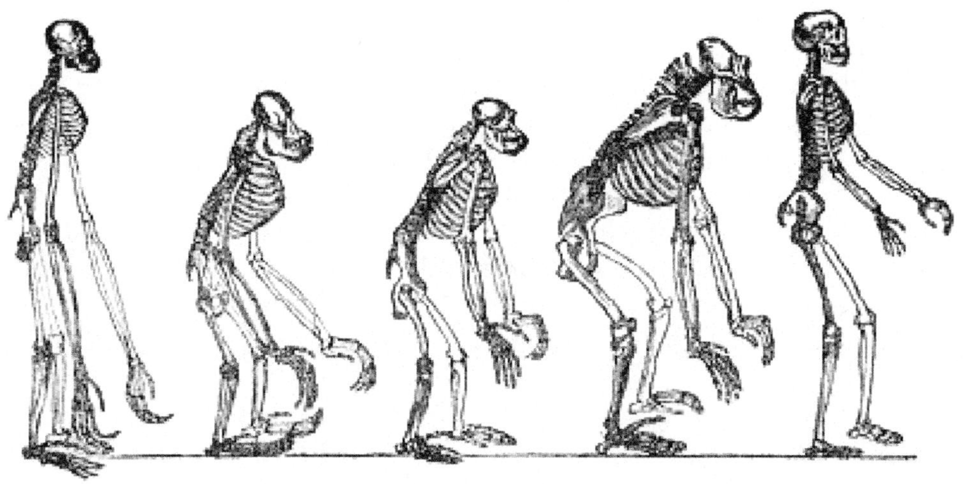

Skeletons of the Gibbon, Orang, Chimpanzee, Gorilla, and Man.
Photographically reduced from diagrams of the natural size (except that
of the Gibbon, which was twice as large as nature), drawn by Mr.
Waterhouse Hawkins from specimens in the Museum of the Royal
College of Surgeons.

…Thus, whatever system of organs be studied, the comparison of their modifications in the ape series leads to one and the same result–that the structural differences which separate Man from the Gorilla and the Chimpanzee are not so great as those which separate the Gorilla from the lower apes.

But in enunciating this important truth I must guard myself against a form of misunderstanding, which is very prevalent. I find, in fact, that those who endeavour to teach what nature so clearly shows us in this matter, are liable to have their opinions misrepresented and their phraseology garbled, until they seem to say that the structural differences between man and even the highest apes are small and insignificant. Let me take this opportunity then of distinctly asserting, on the contrary, that they are great and significant; that every bone of a Gorilla bears marks by which it might be distinguished from the corresponding bone of a Man; and that, in the present creation, at any rate, no intermediate link bridges over the gap between *Homo* and *Troglodytes.*

It would be no less wrong than absurd to deny the existence of this chasm; but it is at least equally wrong and absurd to exaggerate its magnitude, and, resting on the admitted fact of its [145] existence, to refuse to inquire whether it is wide or narrow. Remember, if you will, that there is no existing link between Man and the Gorilla, but do not forget that there is a no less sharp line of demarcation, a no less complete absence of any transitional form, between the Gorilla and the Orang, or the Orang and the Gibbon. I say, not less sharp, though it is somewhat narrower. The structural differences between Man and the Manlike apes certainly justify our regarding him as constituting a family apart from them; though, inasmuch as he differs less from them than they do from other families of the same order, there can be no justification for placing him in a distinct order.

And thus the sagacious foresight of the great lawgiver of systematic zoology, Linnæus, becomes justified, and a century of anatomical research brings us back to his conclusion, that man is a member of the same order (for which the Linnæan term Primates ought to be retained) as the Apes and Lemurs. This order is now divisible into seven families, of about equal systematic value: the first, the Anthropini, contains Man alone; the second, the Catarhini, embraces the old world apes; the third, the Platyrhini, all new world apes, except the Marmosets; the fourth, the Arctopithecini, contains the Marmosets; the fifth, the Lemurini, the Lemurs–from which *Cheiromys* should probably be excluded to form a [146] sixth distinct family, the Cheiromyini; while the seventh, the Galeopithecini, contains only the flying Lemur *Galeopithecus,–* a strange form which almost touches on the Bats, as the *Cheiromys* puts on a Rodent clothing, and the Lemurs simulate Insectivora.

Perhaps no order of mammals presents us with so extraordinary a series of gradations as this–leading us insensibly from the crown and summit of the animal creation down to creatures, from which there is but a step, as it seems, to the lowest, smallest, and least intelligent of the placental Mammalia. It is as if nature herself had foreseen the arrogance of man, and with Roman severity had provided that his intellect by its very triumphs, should call into prominence the slaves, admonishing the conqueror that he is but dust.

These are the chief facts, this the immediate conclusion from them to which I adverted in the commencement of this Essay. The facts, I believe, cannot be disputed; and if so, the conclusion appears to me to be inevitable.

But if Man be separated by no greater structural barrier from the brutes than they are from one another–then it seems to follow that if any process of physical causation can be discovered by which the genera and families of ordinary animals have been produced, that process of causation is [147] amply sufficient to account for the origin of Man. In other words, if it could be shown that the Marmosets, for example, have arisen by gradual modification of the ordinary Platyrhini, or that both Marmosets and Platyrhini are modified ramifications of a primitive stock–then, there would be no rational ground for doubting that man might have originated, in the one case, by the gradual modification of a man-like ape; or, in the other case, as a ramification of the same primitive stock as those apes.

At the present moment, but one such process of physical causation has any evidence in its favour; or, in other words, there is but one hypothesis regarding the origin of species of animals in general which has any scientific existence–that propounded by Mr. Darwin. For Lamarck, sagacious as many of his views were, mingled them with so much that was crude and even absurd, as to neutralize the benefit which his originality might have effected, had he been a more sober and cautious thinker; and though I have heard of the announcement of a formula touching "the ordained continuous becoming of organic forms," it is obvious that it is the first duty of a hypothesis to be intelligible, and that a qua-quâ-versal proposition of this kind, which may be read backwards, or forwards, or sideways, with exactly the same amount of signification, does not really exist, though it may seem to do so.

[148] At the present moment, therefore, the question of the relation of man to the lower animals resolves itself, in the end, into the larger question of the tenability, or untenability, of Mr. Darwin's views. But here we enter upon difficult ground, and it behoves us to define our exact position with the greatest care.

It cannot be doubted, I think, that Mr. Darwin has satisfactorily proved that what he terms selection, or selective modification, must occur, and does occur, in nature; and he has also proved to superfluity that such selection is competent to produce forms as distinct, structurally, as some genera even are. If the animated world presented us with none but structural differences, I should have no hesitation in saying that Mr. Darwin had demonstrated the existence of a true physical cause, amply competent to account for the origin of living species, and of man among the rest.

But, in addition to their structural distinctions, the species of animals and plants, or at least a great number of them, exhibit physiological characters–what are known as distinct species, structurally, being for the most part either altogether incompetent to breed one with another; or if they breed, the resulting mule, or hybrid, is unable to perpetuate its race with another hybrid of the same kind.

A true physical cause is, however, admitted to be such only on one condition–that it shall [149] account for all the phenomena which come within the range of its operation. If it is inconsistent with any one phenomenon, it must be rejected; if it fails to explain any one phenomenon, it is so far weak, so far to be suspected; though it may have a perfect right to claim provisional acceptance.

Now, Mr. Darwin's hypothesis is not, so far as I am aware, inconsistent with any known biological fact; on the contrary, if admitted, the facts of Development, of Comparative Anatomy, of Geographical Distribution, and of Palæontology, become connected together, and exhibit a meaning such as they never possessed before; and I, for one, am fully convinced, that if not precisely true, that hypothesis is as near an approximation to the truth as, for example, the Copernican hypothesis was to the true theory of the planetary motions.

But, for all this, our acceptance of the Darwinian hypothesis must be provisional so long as one link in the chain of evidence is wanting; and so long as all the animals and plants certainly produced by selective breeding from a common stock are fertile, and their progeny are fertile with one another, that link will be wanting. For, so long, selective breeding will not be proved to be competent to do all that is required of it to produce natural species.

I have put this conclusion as strongly as possible before the reader, because the last posi[150]tion in which I wish to find myself is that of an advocate for Mr. Darwin's, or any other views; if by an advocate is meant one whose business it is to smooth over real difficulties, and to persuade where he cannot convince.

In justice to Mr. Darwin, however, it must be admitted that the conditions of fertility and sterility are very ill understood, and that every day's advance in knowledge leads us to regard the hiatus in his evidence as of less and less importance, when set against the multitude of facts which harmonize with, or receive an explanation from, his doctrines.

I adopt Mr. Darwin's hypothesis, therefore, subject to the production of proof that physiological species may be produced by selective breeding; just as a physical philosopher may accept the undulatory theory of light, subject to the proof of the existence of the hypothetical ether; or as the chemist adopts the atomic theory, subject to the proof of the existence of atoms; and for exactly the same reasons, namely, that it has an immense amount of prima facie probability: that it is the only means at present within reach of reducing the chaos of observed facts to order; and lastly, that it is the most powerful instrument of investigation which has been presented to naturalists since the invention of the natural system of classification, and the commencement of the systematic study of embryology.

[151] But even leaving Mr. Darwin's views aside, the whole analogy of natural operations furnishes so complete and crushing an argument against the intervention of any but what are termed secondary causes, in the production of all the phenomena of the universe; that, in view of the intimate relations between Man and the rest of the living world, and between the forces exerted by the latter and all other forces, I can see no excuse for doubting that all are co-ordinated terms of Nature's great progression, from the formless to the formed–from the inorganic to the organic–from blind force to conscious intellect and will.

Science has fulfilled her function when she has ascertained and enunciated truth; and were these pages addressed to men of science only, I should now close this Essay, knowing that my colleagues have learned to respect nothing but evidence, and to believe that their highest duty lies in submitting to it, however it may jar against their inclinations.

But desiring, as I do, to reach the wider circle of the intelligent public, it would be unworthy cowardice were I to ignore the repugnance with which the majority of my readers are likely to meet the conclusions to which the most careful and conscientious study I have been able to give to this matter, has led me.

On all sides I shall hear the cry–"We are men [152] and women, not a mere better sort of apes, a little longer in the leg, more compact in the foot, and bigger in brain than your brutal Chimpanzees and Gorillas. The power of knowledge–the conscience of good and evil–the pitiful tenderness of human affections, raise us out of all real fellowship with the brutes, however closely they may seem to approximate us."

To this I can only reply that the exclamation would be most just and would have my own entire sympathy, if it were only relevant. But, it is not I who seek to base Man's dignity upon his great toe, or insinuate that we are lost if an Ape has a hippocampus minor. On the contrary, I have done my best to sweep away this vanity. I have endeavoured to show that no absolute structural line of demarcation, wider than that between the animals which immediately succeed us in the scale, can be drawn between the animal world and ourselves; and I may add the expression of my belief that the attempt to draw a psychical distinction is equally futile, and that even the highest faculties of feeling and of intellect begin to germinate in lower forms of life. At the same [153] time, no one is more strongly convinced than I am of the vastness of the gulf between civilized man and the brutes; or is more certain that whether *from* them or not, he is assuredly not *of* them. No one is less disposed to think lightly of the present dignity, or despairingly of the future hopes, of the only consciously intelligent denizen of this world.

We are indeed told by those who assume authority in these matters, that the two sets of opinions are incompatible, and that the belief in the unity of origin of man and brutes involves the brutalization and degradation of the former. But is this really so? Could not a sensible child confute by obvious arguments, the shallow rhetoricians who would force this conclusion upon us? Is it, indeed, true, that the Poet, or the Philosopher, or the Artist whose genius is the glory of his age, is degraded from his high estate by the [154] undoubted historical probability, not to say certainty, that he is the direct descendant of some naked and bestial savage, whose intelligence was just sufficient to make him a little more cunning than the Fox, and by so much more dangerous than the Tiger? Or is he bound to howl and grovel on all fours because of the wholly unquestionable fact, that he was once an egg, which no ordinary power of discrimination could distinguish from that of a Dog? Or is the philanthropist, or the saint, to give up his endeavours to lead a noble life, because the simplest study of man's nature reveals, at its foundations, all the selfish passions, and fierce appetites of the merest quadruped? Is mother-love vile because a hen shows it, or fidelity base because dogs possess it?

The common sense of the mass of mankind will answer these questions without a moment's hesitation. Healthy humanity, finding itself hard pressed to escape from real sin and degradation, will leave the brooding over speculative pollution to the cynics and the "righteous overmuch" who, disagreeing in everything else, unite in blind insensibility to the nobleness of the visible world, and in inability to appreciate the grandeur of the place Man occupies therein.

Nay more, thoughtful men, once escaped from the blinding influences of traditional prejudice, will find in the lowly stock whence Man has sprung, the best evidence of the splendour of his [155] capacities; and will discern in his long progress through the Past, a reasonable ground of faith in his attainment of a nobler Future.

They will remember that in comparing civilised man with the animal world, one is as the Alpine traveller, who sees the mountains soaring into the sky and can hardly discern where the deep shadowed crags and roseate peaks end, and where the clouds of heaven begin. Surely the awestruck voyager may be excused if, at first, he refuses to believe the geologist, who tells him that these glorious masses are, after all, the hardened mud of primeval seas, or the cooled slag of subterranean furnaces–of one substance with the dullest clay, but raised by inward forces to that place of proud and seemingly inaccessible glory.

But the geologist is right; and due reflection on his teachings, instead of diminishing our reverence and our wonder, adds all the force of intellectual sublimity to the mere æsthetic intuition of the uninstructed beholder.

And after passion and prejudice have died away, the same result will attend the teachings of the naturalist respecting that great Alps and Andes of the living world–Man. Our reverence for the nobility of manhood will not be lessened by the knowledge that Man is, in substance and in structure, one with the brutes; for, he alone possesses the marvellous endowment of intelligible and rational speech, whereby, in the secular period [156] of his existence, he has slowly accumulated and organised the experience which is almost wholly lost with the cessation of every individual life in other animals; so that, now, he stands raised upon it as on a mountain top, far above the level of his humble fellows, and transfigured from his grosser nature by reflecting, here and there, a ray from the infinite source of truth.

Understanding Human Anatomy through Evolution

Printed in Dunstable, United Kingdom